T0327611

MICROWAVE NONCONTACT MOTION SENSING AND ANALYSIS

WILEY SERIES IN MICROWAVE AND OPTICAL ENGINEERING

KAI CHANG, Editor
Texas A&M University

A complete list of the titles in this series appears at the end of this volume.

MICROWAVE NONCONTACT MOTION SENSING AND ANALYSIS

CHANGZHI LI
Texas Tech University, Lubbock, Texas, USA

JENSHAN LIN
University of Florida, Gainesville, Florida, USA

Published by John Wiley & Sons, Inc., Hoboken, New Jersey
Published simultaneously in Canada

For general information on our other products and services or for technical support, please contact our Customer Care Department within the United States at (800) 762-2974, outside the United States at (317) 572-3993 or fax (317) 572-4002.

Wiley also publishes its books in a variety of electronic formats. Some content that appears in print may not be available in electronic formats. For more information about Wiley products, visit our web site at www.wiley.com.

Library of Congress Cataloging-in-Publication Data:

Li, Changzhi, 1982–
  Microwave noncontact motion sensing and analysis / Changzhi Li, Jenshan Lin.
    pages cm
  Includes index.
  ISBN 978-0-470-64214-6 (hardback)
 1. Motion detectors. 2. Microwave detectors. 3. Motion–Measurement. 4. Radar. I. Lin, Jenshan, 1964– II. Title.
  TK7882.M68L53 2013
  681'.2–dc23

                            2013014155

10 9 8 7 6 5 4 3 2 1

To our families who have been patient with us,
and our colleagues who have been working with us
on this interesting research subject.

CONTENTS

PREFACE

By sending a microwave signal toward a target and analyzing the reflected echo, microwave radar sensors may be used for noncontact motion sensing in many applications. Typical applications range from long distance detection, such as weather radar and automobile speed radar, to short distance monitoring, such as human vital sign detection and mobile tumor tracking. While some speed detection and navigation devices are well known and have been used in the field for decades, other applications based on recent advancements in microwave sensing technologies show great promise and attract interest from practitioners. For example, the principle of detecting motion based on phase shift in a reflected radar signal can be used to sense tiny physiological movements induced by breathing and heartbeat, without any sensor attached to the body. This remote vital sign detection method leads to several potential applications, such as searching for survivors after earthquakes, and monitoring sleeping infants or adults to detect abnormal breathing conditions. Another recent advance in the microwave motion sensing technology is the extension from detecting one-dimensional to two-dimensional rotational movement, which can be used to monitor the spin speed of motors and servos in macroscale machineries and microscale microelectromechanicalsystem (MEMS) devices.

These emerging technologies for health care and industrial sensing provide the advantage of neither confining nor inhibiting the target, as other contact-based technologies do. They enable fast remote identification of hidden target signatures, indicating promising applications in remote diagnosis, search, monitoring, and surveillance. Sensors may be used for the monitoring and treatment of sleep apnea and sudden infant death syndrome. They can outperform other technologies in motion-adaptive tumor tracking during cancer radiotherapy in many anatomic sites. When configured as a nonlinear vibrometer, the radar will also advance approaches to monitoring rotating and reciprocating machinery in the transportation and manufacturing industries. This may positively impact our society through dynamic structural health monitoring, as many buildings worldwide are structurally deficient or functionally obsolete.

While the emerging microwave motion sensing technologies predict attractive ways to replace traditional devices, they generally involve sophisticated hardware and signal analysis that leverage modern design, fabrication, and signal processing methods. Some of the technologies are positioned at the crossroads of electrical engineering, health care and life science, civil engineering, and micro fabrication. With growing interest in multidisciplinary development in the engineering community, researchers and engineers have created various microwave motion sensor front-end architectures and baseband methods.

Although researchers are working on microwave motion sensing technology, and a large number of articles have been published in recent years on state-of-the-art applications, there are not many books that review this technology and unveil the trends of future development. This book aims to review the fundamentals of microwave radar, discusses the state-of-the-art developments, and illustrates future trends.

This book is organized as follows. Chapter 1 introduces the background and recent progress on microwave noncontact motion sensors. Chapter 2 reviews theory and fundamentals of microwave motion sensors. It presents general information about antennas, electromagnetic propagation, link budget, and signal processing. Then it covers typical types of motion sensing radar including Doppler radar, pulse radar,

.and frequency modulated continuous wave (FMCW) radar. Chapter 2 also discusses the key theory and techniques of motion sensing radar. The recent hardware developments of microwave motion sensor are discussed in Chapter 3, including radar transceiver architectures, antenna systems, and special building blocks. In Chapter 4, advances in detection and analysis techniques are discussed, covering system considerations, modeling, simulation, and signal processing. Several application case studies are provided in the first part of Chapter 5, followed by discussion on development of standards and state of acceptance. Finally, future development trends and microwave industry outlook are presented in Chapter 5.

The authors have years of experience working together on microwave noncontact motion sensing technologies from bench-top modules to CMOS integrated microchips, covering a frequency range of over 30 GHz. Besides presenting the history, theory, and technical details of related technologies, the authors provide plenty of application-oriented case studies. Furthermore, the authors exemplify the tight connections of this technology to healthcare, industrial, and military services. Potential research booms are also illustrated to scientists from microwave, electronic circuit, signal processing, and healthcare perspectives.

The authors give their respects to Prof. James Lin at the University of Illinois at Chicago, who pioneered the research of radar noncontact vital sign detection. The authors would like to acknowledge Drs. Arye Rosen, Aly Fathy, and Hao Ling for providing valuable review to the book proposal. Their valuable comments are very much appreciated. The authors would also like to thank their colleagues T.-S. Jason Horng, Olga Boric-Lubecke, Victor Lubecke, Lixin Ran, Jian Li, Wenhsing Wu, Michael Weiss, and Xiaolin "Andy" Li for their valuable discussions and collaborations in the past decade. In addition, the valuable contributions from students who worked diligently on radar motion sensing projects at the University of Florida and Texas Tech University: Yanming Xiao, Changzhan Gu, Te-Yu Jason Kao, Yan Yan, Xiaogang Yu, Gabriel Reyes, Julie Cummings, Jeffrey Lam, and Eric Gravesare are greatly acknowledged. Last but not least, the authors would also like to thank Tien-Yu Huang, Bozorgmehr Vosooghi, and Yiran Li for assistance in preparing figures, indexing, and proofreading the manuscript.

The intended audience of this book includes microwave engineers and researchers, microwave application engineers, researchers in healthcare institutes, developers of military and security equipment, and biomedical engineers.

CHANGZHI LI AND JENSHAN LIN

1

INTRODUCTION

1.1 BACKGROUND

Microwave radar has been used for remote sensing applications for many years. Most common applications include displacement and low velocity measurement (Kim and Nguyen, 2003; Kim and Nguyen, 2004; Benlarbi et al., 1990; Rasshofer and Biebl, 1999), distance and position sensing (Stezer et al., 1999), automobile speed sensing (Meinel, 1995), and vital sign detection (Lin, 1992). Traditionally, microwave radar can be divided into two categories: the pulse radar and the Doppler radar. The pulse radar determines the target range by measuring the round-trip time of a pulsed microwave signal. It does not directly measure the velocity of a target but the velocity can be calculated. The Doppler radar, on the other hand, measures the velocity of a target directly. If the target has a velocity component, the returned signal will be shifted in frequency, due to the Doppler effect. On the hardware side, the pulse radar uses powerful magnetrons to generate microwave signals with very short pulses of applied voltage. In order to overcome the pulse radar's disadvantage of high cost due to the expensive magnetron, the frequency modulated continuous wave

Microwave Noncontact Motion Sensing and Analysis, First Edition.
Changzhi Li and Jenshan Lin.
© 2014 John Wiley & Sons, Inc. Published 2014 by John Wiley & Sons, Inc.

(FMCW) radar was invented in recent years. Compared with the pulse radar, the FMCW radar can be integrated with solid-state technology, and has the advantages of superior target definition, low power, and better clutter rejection. However, the FMCW radar requires accurate control (modulation) of both the frequency and amplitude of the transmitted signal, and is mainly used for range detection. To measure both the displacement and the velocity, a system using millimeter-wave interferometry (Kim and Nguyen, 2003) was reported. It used a quadrature mixer to realize the coherent phase-detection process effectively. It has a very high detection resolution, but has a limit on the minimum measurable speed (Kim and Nguyen, 2004).

1.2 RECENT PROGRESS ON MICROWAVE NONCONTACT MOTION SENSORS

With contributions from many researchers in this field, new detection methods and system architectures have been proposed to improve the detection accuracy and robustness. The advantage of noncontact/covert detection has drawn interest on various applications. While many of the reported systems are bench-top prototypes for concept demonstration, several portable systems and integrated radar chips have been demonstrated.

1.2.1 Microwave/Millimeter-Wave Interferometer and Vibrometer

The development of various instrumentations and techniques for vibration measurement and analysis has become increasingly important. Conventional vibration sensing elements comprise displacement or velocity transducers. One of the most widely used is the accelerometer. A piezoelectric-based accelerometer can produce an electrical output proportional to the vibratory acceleration of the target it is attached to. Another contact measurement instrument is the linear variable differential transformer (LVDT), which works as a displacement transducer that can measure the vibratory displacement directly.

Some of noncontact vibration measurement instruments are laser based, such as laser Doppler vibrometer, laser interferometer, and laser displacement sensor. These devices are usually expensive and have their limitations as well, such as inevitable calibration and narrow detection range. On the other hand, microwave/millimeter-wave interferometer or vibrometer has been used for applications in instrumentation such as plasma diagnostics and nondestructive characterization of material.

Millimeter-wave interferometric sensor with submillimeter resolution has been reported by Kim and Nguyen (2003). Resolving displacement within a fraction of a carrier wavelength, the sensor has high resolution in submillimeter range. The sensor system operates in Ka-band and is completely fabricated using microwave and millimeter-wave integrated circuits. Radio frequency (RF) vibrometer based on nonlinear Doppler phase modulation effect (Li and Lin, 2007a) has also been reported most recently. It detects vibration movement by analyzing the relative strength of vibration-caused harmonics at the radar baseband output. With a quadrature architecture supporting a complex signal demodulation technique, the RF vibrometer realizes the measurement of not only a purely sinusoidal periodic movement, but also vibrations comprised of multiple sine waves of different frequencies. Compared with laser-based sensors, microwave/millimeter-wave interferometer and vibrometer can have a low cost and a much larger detection range.

1.2.2 Noncontact Vital Sign Detection

The principle of detection based on frequency or phase shift in a reflected radar signal can be used to detect tiny body movements induced by breathing and heartbeat, without any sensor attached to the body. There are several advantages to a noncontact vital sign detection solution: physically, it neither confines nor inhibits the subject, making the detector ideal for long-term continuous monitoring applications. Also, the reliability can be increased as a subject is unaware of the measurement and therefore is less likely to alter their vital signs. Additionally, accuracy is enhanced because of the lack of surface loading effects that have been shown to reduce the accuracy of some other measurement methods. This noncontact remote detection of vital signs leads to several potential applications such as searching for survivors after an earthquake and monitoring sleeping infants or adults to detect abnormal breathing conditions.

While the concept of noncontact detection of vital signs has been successfully demonstrated by pioneers in this field before 2000 (Lin, 1975; Chuang et al., 1991; Lin, 1992; Chen 1986), research efforts in this century have been moving the technology development toward lower power, lighter weight, smaller form factor, better accuracy, longer detection range, and more robust operation for portable and handheld applications. Among many possible applications this technology can be used for, healthcare seems to be drawing most of the interest. As an example, a baby monitor using this technology was recently demonstrated (Li et al., 2009a). The baby monitor integrates a low power Doppler radar to detect tiny baby movements induced by breathing. If no movement is detected within 20 s, an alarm will be triggered. With growing interests in health and life sciences by the engineering community, many researchers have been contributing to technology advancement in this field. Although, many results were demonstrated using bench-top prototypes or board-level integration, their architectures still show the potential of being implemented on chip. In fact, there have been several reports of vital sign radar sensor chips based on various architectures (Droitcour et al., 2002; Droitcour et al., 2003; Li et al., 2008b; Li et al., 2009c; Li et al., 2010b).

1.3   ABOUT THIS BOOK

Although, many researchers are working on the microwave motion sensing technology and a large number of articles have been published in recent years on state-of-the-art applications such as vital sign detection and interferometry, it is difficult to find a book that reviews this technology and unveils the trends of future development. This book first reviews the theory and fundamentals of microwave motion sensor in Chapter 2. It then discusses the hardware development of microwave motion sensor in Chapter 3, including radar transceiver architectures, antenna systems, and special building blocks. In Chapter 4, advances in detection and analysis techniques will be discussed, covering system consideration, modeling, and signal processing. Several application case studies will be provided in the first part of Chapter 5, followed by the discussion on development of standards and state of acceptance. Finally, future development trends and microwave industry outlook will be presented in Chapter 5.

This book not only covers the theory and technical details of related technologies, but also plenty of applications. The tight connections of this technology to healthcare, industrial, and military services will be exemplified in this book. Potential research opportunities will also be illustrated to scientists from the microwave, electronic circuit, signal processing, and healthcare points of view. The intended audience of this book includes microwave engineers and researchers, microwave application engineers, researchers in healthcare institutes, developers of military and security equipment, and scientists in biomedical engineering.

2

THEORY OF MICROWAVE NONCONTACT MOTION SENSORS

2.1 INTRODUCTION TO RADAR

The word "radar" originally stands for *radio detection and ranging*. It is so commonly used today and the word has become a standard English noun. Although, it is often conceived as a technology developed during World War II, the history of radar actually extends back to the time well before World War II when researchers in many countries started the research that led to the development of radar (Page, 1962; Shipton, 1980; James, 1989; Chernyak and Immoreev, 2009; Guarnieri, 2010). In 1904, the first patent on the detection of objects using radio waves was given to the German engineer Hülsmeyer, who experimented with target detection by bouncing waves off a ship. The device was called *Telemobiloskop* by Hülsmeyer. In 1922, Marconi advocated this idea again. In the same year, Taylor and Young of the US Naval Research Laboratory demonstrated ship detection by radar, which was based on an interference pattern when a ship passed between transmitting and receiving antennas. In 1930, Hyland, a colleague of Taylor and Young, first detected an aircraft by radar, setting

Microwave Noncontact Motion Sensing and Analysis, First Edition.
Changzhi Li and Jenshan Lin.
© 2014 John Wiley & Sons, Inc. Published 2014 by John Wiley & Sons, Inc.

off a more substantial investigation that led to a US patent for today's continuous wave (CW) radar in 1934. After that, in the middle and late 1930s, largely independent developments of radar accelerated and spread in countries including the United States, the Soviet Union, the United Kingdom, Germany, France, Japan, Italy, and the Netherlands (Swords, 1986; Watson, 2009; Richards, 2005).

Military applications, including surveillance, navigation, and weapons guidance for ground, sea, and air vehicles, were the early driving force for radar development. In recent years, however, radar has started to enjoy an increasing range of applications in civilian life. One of the most common applications is the police traffic radar used to enforce speed limits. This technology is also used to measure the speed of baseballs and tennis serves. Other civilian applications include automotive collision avoidance radar, Doppler weather radar, air traffic control system for guiding commercial aircrafts, and aviation radar. Also, spaceborne or airborne radar is an important tool in mapping earth topology and environmental characteristics such as water and ice conditions, forestry conditions, land usage, and pollution.

In this section, radar basics will be reviewed. The review starts with antenna, which is followed by wave propagation, radio system link, Friis equation, radar cross section, radar equation, and radar signal-to-noise ratio (SNR). Finally, radar signal processing basics will be briefly introduced.

2.1.1   Antennas

As an indispensable building component of a radar system, an antenna is a device for radiating or receiving radio waves. It is a transitional structure between free space and a guiding device such as transmission line. It converts radiated waves into guided waves, or vice versa. Antennas are inherently bidirectional, in that they can be used for both transmitting and receiving functions. Examples of antennas include dipole antenna, monopole antenna, patch antenna, horn antenna, reflector antenna (such as dish and parabolic antenna), and phased-array antenna (Balanis, 2005).

Antenna radiation pattern is generally used to describe the radiation of electromagnetic waves into the space. It is a graphical representation of the radiation intensity of the antenna as a function of space coordinates (in most cases, directional coordinates). In system design and calculation, the isotropic antenna, which is a hypothetical antenna having equal radiation in all directions, is usually used as a reference. In contrast, a directional antenna refers to an antenna that radiates or receives signals more effectively in some directions than in others. One terminology often used is the omnidirectional antenna which is a type of antenna having a nondirectional pattern in azimuth and a directional pattern in elevation. A dipole antenna is an example of omnidirectional antenna.

Antennas radiate or receive electromagnetic wave. Therefore, the E-plane pattern is defined as the radiation pattern in a plane containing the electric field vector and the direction of the maximum radiation. On the other hand, the H-plane pattern is defined as the radiation pattern in a plane containing the magnetic field vector and the direction of the maximum radiation. As an example, Figure 2.1 shows E-plane and H-plane radiation patterns for a dipole antenna.

A fundamental property of an antenna is its ability to focus power in a given direction, to the exclusion of other directions. An antenna with a broad main beam can transmit or receive power over a wide angular region, while an antenna having a narrow main beam will transmit or receive power over a small angular region. A measure of this focusing ability is the half-power beamwidth or the 3-dB beamwidth of the antenna, which is the angular width defined by two half-power points of the main beam in a two-dimensional plane. At the half-power point,

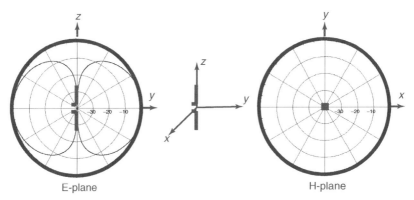

Figure 2.1   E-plane and H-plane radiation patterns of a dipole antenna.

the radiation intensity in far field drops by half (about -3 dB) from its maximum value. Another measure in three-dimensional form is the antenna beam solid angle which is defined as the solid angle of an equivalent radiation pattern that has constant radiation intensity equal to the maximum radiation intensity of the actual radiation pattern and contains all the radiated power.

Another two terms often used are the far field and the near field. The far field is defined as the distance where the spherical wave front radiated by an antenna becomes a close approximation to the ideal planar phase front of a plane wave. In the far-field region, angular field distribution (antenna pattern) is essentially independent of the distance from the antenna. This approximation applies over the aperture area of the antenna, and thus depends on the maximum dimension of the antenna, denoted as D. When D is larger than the free-space wavelength λ, the far-field distance is defined as $r > 2D^2/\lambda$. When D is close to λ or smaller than λ, the far-field distance should be at $r \gg D$.

2.1.2 Propagation and Antenna Gain

Consider an antenna located at the origin of a spherical coordinate system. At large distances where the local near-zone fields are negligible, the radiated electric field of an arbitrary antenna can be expressed as

$$\overrightarrow{E}(r, \theta, \phi) = [\widehat{\theta} F_\theta(\theta, \phi) + \widehat{\phi} F_\phi(\theta, \phi)]\frac{e^{-jk_0 r}}{r}, \qquad (2.1)$$

where $\overrightarrow{E}$ is the electric field vector with a unit of voltage over distance (e.g., V/m), $\widehat{\theta}$ and $\widehat{\phi}$ are unit vectors in the spherical coordinate system, r is the radial distance from the origin, $k_0 = 2\pi/\lambda$ is the free-space propagation constant, with wavelength $\lambda = c/f$, $F_\theta(\theta, \phi)$, and $F_\phi(\theta, \phi)$ are the pattern functions. Equation 2.1 describes an electric field propagating in the radial direction, with a phase variation of $e^{-jk_0 r}$ and an amplitude variation of $1/r$. The electric field may be polarized in either the $\widehat{\theta}$ or $\widehat{\phi}$ directions, but not in the radial direction, as this is a transverse electromagnetic (TEM) wave. The electric and magnetic fields can be correlated using $H_\phi = E_\theta/\eta_0$ and $H_\theta = -E_\phi/\eta_0$, where $\eta_0 = 377 \ \Omega$ is the wave impedance in free space. The Poynting vector for this wave, which represents the energy flux, is given by

$$\vec{S} = \vec{E} \times \vec{H}^* \ \text{W/m}^2,\tag{2.2}$$

and the radiation density, which is the time average of Poynting vector having a unit of power over area (e.g., W/m^2), can be obtained as

$$W_{\text{rad}} = \vec{S}_{\text{avg}} = \frac{1}{2}\text{Re}\{\vec{S}\} = \frac{1}{2}\text{Re}\{\vec{E} \times \vec{H}^*\} \ \text{W/m}^2.\tag{2.3}$$

On the basis of radiation density, we can define the radiation intensity of the radiated electromagnetic field as

$$U(\theta, \phi) = r^2 \times W_{\text{rad}} = \frac{r^2}{2}\text{Re}\{E_\theta \hat{\theta} \times H_\phi^* \hat{\phi} + E_\phi \hat{\phi} \times H_\theta^* \hat{\theta}\}$$

$$= \frac{r^2}{2\eta_0}[|E_\theta|^2 + |E_\phi|^2]$$

$$= \frac{1}{2\eta_0}[|F_\theta|^2 + |F_\phi|^2] \ \text{W/unit solid angle.}\tag{2.4}$$

Note that the unit of the radiation intensity is power per unit solid angle or power per steradian (e.g., W/sr), as the radial dependence has been removed. The radiation intensity gives the variation in radiated power versus direction from the antenna. The total power radiated by the antenna can be found by integrating the Poynting vector over the surface of a sphere of radius r that encloses the antenna. This is equivalent to integrating the radiation intensity over a unit sphere:

$$P_{\text{rad}} = \int_{\phi=0}^{2\pi}\int_{\theta=0}^{\pi} \vec{S}_{\text{avg}} \cdot \hat{r} r^2 \sin\theta \, d\theta \, d\phi = \int_{\phi=0}^{2\pi}\int_{\theta=0}^{\pi} U(\theta, \phi) \sin\theta \, d\theta \, d\phi.\tag{2.5}$$

For an arbitrary antenna, the average radiation intensity is

$$U_{\text{avg}} = r^2 \times \frac{P_{\text{rad}}}{4\pi \times r^2} = \frac{P_{\text{rad}}}{4\pi}.\tag{2.6}$$

The effective isotropic radiated power EIRP is defined using $U(\theta,\phi)$ $= \frac{\text{EIRP}}{4\pi}$, which leads to

$$\text{EIRP} = 4\pi \times U(\theta, \phi) = 4\pi \times r^2 \times W_{\text{rad}}. \qquad (2.7)$$

A measure of the focusing ability of an antenna is the directivity, defined as the ratio of the maximum radiation intensity in the main beam to the average radiation intensity

$$D = \frac{U_{\text{max}}}{U_{\text{avg}}} = \frac{4\pi \times U_{\text{max}}}{P_{\text{rad}}}. \qquad (2.8)$$

Directivity is a dimensionless ratio of power, and is usually expressed in decibel as $D(\text{dB}) = 10\log(D)$.

The radiation intensity of an isotropic antenna is constant at all angles for any value of θ and ϕ:

$$U_0 = r^2 \times \frac{P_{\text{rad}}}{4\pi \times r^2} = \frac{P_{\text{rad}}}{4\pi}. \qquad (2.9)$$

By applying this to the definition of the antenna directivity, it is shown that the directivity of an isotropic element is $D = 1$, or 0 dB. As the minimum directivity of any antenna is unity, the directivity is sometimes stated as relative to the directivity of an isotropic radiator, and written as dBi. Examples of directivities of some typical antennas are 2.15 dB for a half-wavelength dipole antenna and 1.5 dB for a short dipole antenna.

The mismatch between the transmission line feeding the antenna and the antenna input port causes reflection loss to the signal power. Also, resistive losses, due to nonideal metals and dielectric materials, exist in all antennas. Such reflective losses and resistive losses result in a difference between the power delivered to the input of an antenna and the power radiated by that antenna. Therefore, we can define the radiation efficiency e_t of an antenna as the ratio of the total radiated power to the supplied input power:

$$e_t = \frac{P_{\text{rad}}}{P_{\text{in}}} = 1 - \frac{P_{\text{loss}}}{P_{\text{in}}} = e_r \times e_c \times e_d = (1 - |\Gamma|^2) \times e_{cd}, \qquad (2.10)$$

where e_r is the reflection efficiency, e_c is the conduction efficiency, e_d is the dielectric efficiency, e_{cd} is the conduction–dielectric efficiency combined, and Γ is the reflection coefficient at the antenna input port.

Recall that antenna directivity is a function of only the shape of the antenna radiation pattern (i.e., the radiated fields), and is not affected by losses in the antenna. In order to account for the fact that an antenna with radiation efficiency less than unity will not radiate all of its input power, we define antenna gain as the ratio of the EIRP to the input power. It can be calculated as the product of directivity and efficiency

$$G = e_t D = \frac{\text{EIRP}}{P_{\text{in}}} = \frac{4\pi \times U(\theta, \phi)}{P_{\text{in}}}. \tag{2.11}$$

From Equation 2.11, it is clear that antenna gain is always less than or equal to directivity. Just like the antenna directivity, antenna gain is usually expressed in decibel, as $G(\text{dB}) = 10 \log(G)$.

2.1.3 Radio System Link and Friis Equation

A general radio system link is shown in Fig. 2.2, where the transmit power is P_t, the transmit antenna gain is G_t, the receive antenna gain is G_r, and the received power delivered to a matched load is P_r. The transmit and receive antennas are separated by a distance r.

From the previous discussion, the power density radiated by an isotropic antenna ($D = 1 = 0$ dB) at a distance r is given by

$$S_{\text{avg}} = \frac{P_t}{4\pi r^2} \ \text{W/m}^2. \tag{2.12}$$

This result reflects the fact that we must be able to recover all of the radiated power by integrating over a sphere of radius r surrounding the antenna, as the power is distributed isotropically, and the area of

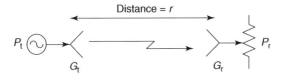

Figure 2.2   A general radio system link.

a sphere is $4\pi r^2$. If the transmit antenna has a directivity greater than 0 dB, we can find the radiated power density by multiplying Equation 2.12 and the directivity, as the directivity is defined as the ratio of the actual radiation intensity to the equivalent isotropic radiation intensity. Also, if the transmit antenna has losses, we can include the radiation efficiency factor, which converts directivity to gain. Thus, the general expression for the power density radiated by an arbitrary transmit antenna is

$$S_{avg} = \frac{G_t P_t}{4\pi r^2} \text{ W/m}^2. \tag{2.13}$$

With this power density at the receive antenna, we can use the concept of effective aperture area to find the received power:

$$P_r = A_e S_{avg} = \frac{G_t P_t A_e}{4\pi r^2} \text{ W}, \tag{2.14}$$

where the effective aperture area has dimensions of square meter, and can be interpreted as the capture area of a receive antenna. It can be shown that the maximum effective aperture area of an antenna is related to the directivity of the same antenna as

$$A_e = \frac{D\lambda^2}{4\pi}, \tag{2.15}$$

where λ is the operating wavelength of the antenna. The maximum effective aperture area as defined above does not include the effect of losses in the antenna, which can be accounted for by replacing D with the gain of the antenna G.

Using the relationship between the antenna maximum effective aperture area and gain, the result for the received power is

$$P_r = \frac{G_t G_r \lambda^2}{(4\pi r)^2} P_t \text{ W}. \tag{2.16}$$

This result is known as the *Friis radio link formula* or *Friis transmission equation*, and it addresses the fundamental question of how much power is received by a receiving antenna from a transmitting antenna located at a certain distance away with a transmitted signal of wavelength λ. In practice, the value given by Equation 2.16 should be interpreted as the ideal received power in free space, as there are a number of factors such as gaseous absorption, rain attenuation, and multipath reflection that can serve to decrease or increase the received power in an actual radio system.

2.1.4 Radar Cross Section and Radar Equation

Radar systems can be divided into two categories: the monostatic radar uses the same antenna for both transmission and reception, while the bistatic radar uses two separate antennas for these functions. Most radars are of the monostatic type, but in some applications (such as missile fire control), the target is illuminated by a separate transmit antenna. Separate antennas are also sometimes used to achieve the necessary isolation between transmitter and receiver. In the following discussion, the transmitting antenna and the receiving antenna are assumed to be the same antenna or collocated.

If the transmitter radiates a power P_t through an antenna of gain G, the incident power density on the target is

$$S_t = \frac{P_t G}{4\pi r^2},\tag{2.17}$$

where r is the distance to the target. It is assumed that the target is in the main beam direction of the antenna. The target will scatter the incident power in various directions. The radar cross section, denoted as σ, is defined as the equivalent area that intercepts incident wave and scatters isotropically to produce at the receiver the same power density scattered by the actual target. In other words, it is the ratio of the effective isotropic scattered power (scattered power by the target calculated from measured power density at radar receiver, assuming isotropic scattering by the target) to the incident power density. Therefore,

$$\frac{\sigma S_t}{4\pi r^2} = S_r,\tag{2.18}$$

where S_r is the power density of the scattered wave at the receiving antenna. The radar cross section has the dimension of area, and is a property of the target itself. It depends on the incident and reflection angles, as well as the polarization of the incident wave. It should be noted that the definition of radar cross section and the above equations are only valid in far field.

Substituting Equation 2.17 into Equation 2.18, Equation 2.18 becomes

$$S_r = \frac{P_t G \sigma}{(4\pi r^2)^2}. \tag{2.19}$$

Using the relationship between the antenna effective area and gain, the received power can be represented as

$$P_r = \frac{G^2 \lambda^2 \sigma}{(4\pi)^3 r^4} P_t. \tag{2.20}$$

This is the radar equation or radar range equation for monostatic radar. Note that the received power varies as $1/r^4$. The received power will be reduced by 12 dB once the distance between radar and target doubles.

2.1.5   Radar Signal-To-Noise Ratio

On the basis of the radar range equation, the SNR of the radar output can be represented as

$$\text{SNR} = \frac{P_r}{P_n} = \frac{P_t G^2 \lambda^2 \sigma}{(4\pi)^3 r^4 kTB F_n L}, \tag{2.21}$$

where P_r is the radar detected signal power usually measured at the output of the matched filter or the signal processor, $P_n = kTB F_n$ is the noise power at the same point that P_r is specified, r is the range from the radar to the target, k is Boltzmann's constant and is equal to 1.38×10^{23} W/(Hz K), T denotes the temperature, B is the effective noise bandwidth of the radar, F_n is the radar noise figure, L is a term included to account for all losses that must be considered when using the radar range equation.

One of the important uses of the radar range equation is in the determination of detection range, or the maximum range at which a target has a high probability of being detected by the radar. Because of the noise received by the antenna and generated in the receiver, there will be some minimum detectable power that can be discriminated by the receiver. If this power is P_{min}, then the maximum radar operation range based on the radar equation is

$$r_{max} = \left[\frac{P_t G^2 \lambda^2 \sigma}{(4\pi)^3 P_{min}} \right]^{1/4}. \tag{2.22}$$

If the SNR threshold value required to detect a target is given as SNR_{min}, the maximum radar detection range can be expressed as

$$r_{\text{max}} = \left[\frac{P_t G^2 \lambda^2 \sigma}{(4\pi)^3 (\text{SNR}_{\text{min}}) kTB\, F_n L} \right]^{1/4}. \qquad (2.23)$$

Signal processing can effectively reduce the minimum detectable signal, and so increase the usable range. For example, a common processing technique used with pulse radars is pulse integration, where a sequence of N received pulses is integrated over time. The effect is to reduce the noise level which has a zero mean relative to the returned pulse level, hence resulting in an improvement factor of approximately N.

2.1.6 Signal-Processing Basics

Most radar applications can be classified as detection, tracking, or imaging. To evaluate the performance of a radar system, a variety of figures of merit can be used, depending on specific applications. In analyzing detection performance, the fundamental parameters are the probability of detection, the probability of false alarm, the signal-to-interference ratio (SIR), resolution, and the detection range. In radar tracking, the basic figure of merit is accuracy of range, angle, and velocity estimation. In imaging, the principal figures of merit are spatial resolution and dynamic range. Spatial resolution determines what object size can be identified in the final image, while dynamic range determines image contrast.

Radar signal processing serves to improve the above figures of merit. SIR can be improved by pulse integration. Resolution and SIR can be jointly improved by pulse compression and other waveform design techniques, such as multiple-input multiple-output (MIMO) waveform synthesis and frequency agility. Increased SIR and filter splitting interpolation methods can benefit accuracy. Side lobe behavior can be improved with the same windowing techniques that are used in virtually every application of signal processing.

Radar signal processing benefits from similar techniques and concepts used in other signal processing areas. They include communications, sonar, and speech and image processing. Statistical detection and linear filtering theory are important to radar's task of target detection. Fourier transforms are used for applications including Doppler spectrum estimation, fast convolution implementations of matched filters, and radar imaging. Modern model-based spectral estimation and adaptive filtering techniques are used for beamforming and jammer cancellation. Pattern recognition techniques are used for target/clutter discrimination and target identification.

In the meantime, radar signal processing has several unique properties that differentiate it from most of the other signal processing fields. Radar signals have very high dynamic ranges of tens of decibels. In some cases, it approaches 100 dB. Thus, gain control schemes are common, and side lobe control is often critical to avoid having weak signals masked by stronger ones. Also, the SIRs are often relatively low. In many applications, radar detected signal bandwidth may be large and thus requires very fast analog-to-digital (A/D) converters (ADCs). Consequently, it is necessary to design custom hardware for the digital signal processor in order to obtain adequate throughput.

2.2   MECHANISM OF MOTION SENSING RADAR

In this section, the fundamental mechanism of motion sensing radar will be introduced. The general types of motion sensing radar that will be discussed include Doppler radar based on the Doppler frequency shift, Doppler radar based on nonlinear phase modulation and demodulation, pulse radar, and FMCW radar. A comparison among these radar types will be conducted at the end of this section using the detection of sinusoidal mechanical vibration as an example.

2.2.1   Doppler Frequency Shift

Doppler frequency shift is the fundamental mechanism for Doppler radar. Figure 2.3 shows a basic Doppler radar system detecting a moving target.

In the basic operation of a Doppler radar, a transmitter sends out a signal $T(t) = \cos(2\pi f_0 t + \Phi_1)$ with carrier frequency f_0 and residual phase Φ_1. The transmitted signal will be partly reflected by a distant target and then detected by a sensitive receiver. If the target has a velocity component v along the line-of-sight of the radar, the returned signal will be shifted in frequency relative to the transmitted

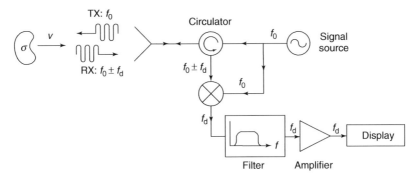

Figure 2.3   Doppler radar system. Adapted from Pozar.

frequency, because of the Doppler Effect. The received signal can be represented as

$$R(t) = \cos \left[2\pi f_0 t \pm \frac{4\pi v t}{\lambda} + \Phi_2 \right], \qquad (2.24)$$

where Φ_1 is changed to Φ_2 because of nominal detection distance d_0 and accumulated phase noise, the plus sign corresponds to an approaching target and the minus sign corresponds to a receding target. Therefore, the shift in frequency, or the Doppler frequency, will be

$$f_d = \frac{2v f_0}{c}, \qquad (2.25)$$

where c is the velocity of light and $\lambda = c/f_0$. The received frequency is then $f_0 \pm f_d$. If the received signal is down-converted by the same transmitted signal, the base band output will be $B(t) = \cos[\pm 2\pi f_d t + \Delta \Phi] = \cos[\pm 4\pi v t/\lambda + \Delta \Phi]$. By detecting the Doppler frequency shift, the target speed can be obtained.

2.2.2   Doppler Nonlinear Phase Modulation

A recent innovation on the Doppler radar technology is the application of the nonlinear Doppler phase modulation effect. The conventional Doppler radar monitors the Doppler frequency shift and determines the speed of a one-directional movement or the frequency of a periodic movement. Taking advantage of the nonlinear phase modulation effect, a nonlinear Doppler radar can be used to attain additional information such as the amplitude of periodic movement (Li and Lin, 2007a). With appropriate software demodulation algorithm, complex movement pattern can possibly be recovered based on the harmonic information. Therefore, the nonlinear Doppler radar will find its advantages in the monitoring of mechanical vibration and rotation.

Figure 2.4 shows the nonlinear Doppler radar detecting a periodic movement $x(t)$. Similar to the conventional Doppler radar, an unmodulated signal $T(t) = \cos(2\pi f_0 t + \Phi_1)$ with carrier frequency f_0 and residual phase Φ_1 is transmitted. Upon reaching the target, the signal is phase modulated by the target movement $x(t)$. The reflected signal is represented as $R(t) = \cos[2\pi ft - 4\pi x(t)/\lambda + \Phi_2]$, where Φ_1 is changed to Φ_2 because of nominal detection distance d_0 and phase noise. Using the same transmitted signal $T(t)$ as the local oscillator (LO) signal, the radar receiver down-converts $R(t)$ into baseband. The normalized radar baseband output signal can be approximated as

$$B(t) = \cos\left(\theta + \frac{4\pi x(t)}{\lambda} + \Delta\Phi\right), \tag{2.26}$$

where θ is a constant phase shift created on the transmission path, $\Delta\Phi$ is the total residual phase noise, and λ is the carrier wavelength. Because of the nonlinear property of the cosine transfer function (or exponential function in a complex system), the baseband output will have a waveform that is not linearly proportional to the original movement $x(t)$.

2.2.2.1 Doppler Phase Modulation with Small-Angle Approximation.
When the movement amplitude is much smaller than the wavelength λ, a linear approximation can be applied (Droitcour et al., 2004a). If θ is an odd multiple of $\pi/2$, the baseband output is approximately,

$$B(t) \approx \frac{4\pi x(t)}{\lambda} + \Delta\Phi. \tag{2.27}$$

In this case, the baseband output is proportional to the motion displacement $x(t)$ and the optimum phase-demodulation sensitivity is achieved.

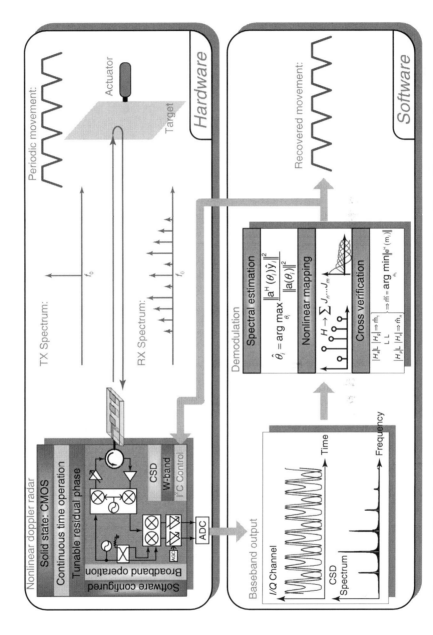

Figure 2.4 Detection of a periodic movement using the proposed nonlinear Doppler radar vibrometer.

If θ is an even multiple of $\pi/2$, the baseband output is approximately,

$$B(t) \approx 1 - \left[\frac{4\pi x(t)}{\lambda} + \Delta\Phi\right]^2. \qquad (2.28)$$

In this case, the baseband output is no longer linearly proportional to the time-varying displacement $x(t)$, and the detection happens at a null detection point. This null point occurs when the LO and received signal $R(t)$ are either in-phase or $180°$ out of phase. As the variable part of θ is dependent only on the distance to the target d_0, the null detection point occurs with a target distance every $\lambda/4$ from the radar. Droitcour et al. proposed to avoid these null detection points with a quadrature receiver, where two receiver chains with LO phases $90°$ apart insure that there is always at least one output not in the null detection point.

The two output channels of a quadrature receiver will be

$$\begin{cases} B_I(t) = \cos\left(\theta + \frac{\pi}{4} + \frac{4\pi x(t)}{\lambda} + \Delta\Phi\right) \\ B_Q(t) = \cos\left(\theta - \frac{\pi}{4} + \frac{4\pi x(t)}{\lambda} + \Delta\Phi\right) \end{cases}. \qquad (2.29)$$

When $\theta + \pi/4$ is an even/odd multiple of $\pi/2$, the I-channel signal will be at the null/optimum detection point while the Q-channel signal will be at the optimum/null detection point. In these cases, there will always be one channel at the optimum detection point to guarantee good detection accuracy. With a quadrature receiver, the worst case is when θ is an integer multiple of π so that both $\theta + \pi/4$ and $\theta - \pi/4$ are odd multiples of $\pi/4$, and neither receiver chain is at the optimum phase-demodulation point. The baseband outputs at this point are

$$B_I(t) = B_Q(t) \approx \frac{1}{\sqrt{2}} - \frac{1}{\sqrt{2}}\left[\left(\frac{4\pi x(t)}{\lambda} + \Delta\Phi\right)\right.$$
$$\left. + \frac{1}{2}\left(\frac{4\pi x(t)}{\lambda} + \Delta\Phi\right)^2\right]. \qquad (2.30)$$

As long as $x(t)$ is much less than λ, the linear term is much larger than the squared term carrying the movement information $x(t)$. Therefore, the fundamental of $x(t)$ will dominate and the motion information can still be detected. Furthermore, it will be shown in Section 2.3 that when complex signal demodulation is used to combine the I-and the Q-channel outputs, robust automatic detection without any requirement of selecting between the I/Q channels can be achieved.

2.2.2.2 *Doppler Phase Modulation Nonlinear Analysis.*

When small wavelength is used such that $x(t)$ is comparable to λ, a series of harmonic frequency tones will be produced. For a single-tone periodic movement, $x(t) = m \sin \omega t$, where m is the movement amplitude, $\omega = 2\pi f_{\mathrm{m}}$ describes the movement frequency. For a more complex movement, it can be decomposed into a series of single-tone movements. Using Fourier series to expand the baseband signal $B(t)$ described by Equation 2.26, it can be expressed as

$$B(t) = 2 \cdot \sum_{k=1}^{\infty} J_{2k}\left(\frac{4\pi m}{\lambda}\right) \cdot \cos 2k\omega t \cdot \cos \Phi$$
$$- 2 \cdot \sum_{k=0}^{\infty} J_{2k+1}\left(\frac{4\pi m}{\lambda}\right) \cdot \sin(2k+1)\omega t \cdot \sin \Phi, \qquad (2.31)$$

where $\Phi = \theta + \Delta\Phi(t)$ is the total residual phase, and $J_n(x)$ is the nth-order Bessel function of the first kind. Therefore, the phase-modulated baseband signal is decomposed into a number of harmonics of the fundamental frequency. While the movement frequency ω is readily obtained from the fundamental frequency of $B(t)$, Equation 2.31 also shows that for a certain carrier frequency, the relative strength among the harmonics is decided by the movement amplitude m and the residual phase Φ, but is not a function of signal level determined by receiver gain and measurement distance. For example, the absolute value of ratio among the first-, second-, third-, and fourth-order harmonics is

$$H_1 : H_2 : H_3 : H_4 = \left|J_1\left(\frac{4\pi m}{\lambda}\right)\cos \Phi\right| : \left|J_2\left(\frac{4\pi m}{\lambda}\right)\sin \Phi\right| :$$
$$\left|J_3\left(\frac{4\pi m}{\lambda}\right)\cos \Phi\right| : \left|J_4\left(\frac{4\pi m}{\lambda}\right)\sin \Phi\right|. \qquad (2.32)$$

Moreover, if separating the harmonics into groups of even order and odd order, the ratio inside each group is only decided by m.

Therefore, m can be found by fitting the measured harmonic ratio to the theoretical value from Bessel function, which can be performed for either odd-order or even-order harmonics. This analysis leads to a very important application—the amplitude of the movement can be accurately determined in remote noncontact measurement without calibrating the signal level versus distance, provided that the wavelength is accurately determined. If linear modulation method is used, the signal level that is affected by the receiver gain and the distance to the target would need to be calibrated to determine the movement amplitude.

If both the even-order and the odd-order harmonic ratios are used to extract the amplitude of movement, there will be more than a single result for the same measurement. The accuracy of the measurement can thus be verified by checking whether the two results agree with each other. Therefore, a pair of harmonic ratios (i.e., an even-order ratio and an odd-order ratio) will be used simultaneously for a single measurement. For example, Fig. 2.5 shows up to the sixth-to-fourth harmonic ratio as a function of normalized movement amplitude. The plot shows that when the displacement is too small compared with the wavelength, the harmonics are too weak to be observed, corresponding to the linear approximation region in Equation 2.27. As the displacement increases, harmonics become observable, making the invented measurement method possible.

It should be noted that there are multi solutions of movement amplitude because of the nonlinear property, and it is impractical to accurately measure harmonic ratios that are either too small or too large. Therefore, a detection range is defined for a pair of harmonic ratios as the lowest range of movement amplitude producing measurable harmonic ratio within the range of 0.2–5 from experience. For example, in Fig. 2.5, the detection range of the third-to-first and the fourth-to-second harmonic ratio pair is 0.214λ–0.290λ, which corresponds to 0.2 for H_4/H_2 and 5 for H_3/H_1, respectively. Similarly, the detection range of the fifth-to-third and the sixth-to-fourth harmonic ratio pair in Fig. 2.5 is 0.335λ–0.489λ. On the basis of the observation above, the applicable detection range, normalized to λ, for different harmonic pairs is calculated in Table 2.1. The H-Pair with index "i" means the $(i + 2)$th-to-ith and the $(i + 3)$th-to-$(i + 1)$th harmonic ratio pair.

From the above table, it is shown that except for a small gap between the $i = 1$ pair and the $i = 2$ pair, the nonlinear detection method can detect any movement amplitude that is larger than 0.214λ.

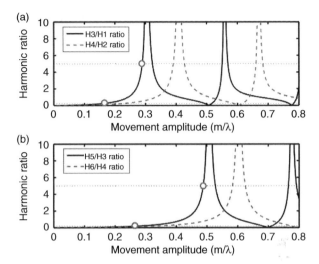

Figure 2.5 Theoretical harmonic ratio as a function of the movement amplitude. The movement amplitude m is normalized to the carrier wavelength λ.
(a) H_3/H_1 and H_4/H_2 ratio. (b) H_5/H_3 and H_6/H_4 ratio (Li and Lin, 2007a).

TABLE 2.1 Detection Range for Different Pairs of Harmonics (m/λ)

H-Pair	$i = 1$	$i = 2$	$i = 3$	$i = 4$	$i = 5$
Lower bound	0.214	0.335	0.455	0.575	0.694
Upper bound	0.290	0.489	0.677	0.859	1.039
H-pair	$i = 6$	$i = 7$	$i = 8$	$i = 9$	$i = 10$
Lower bound	0.813	0.932	1.051	1.170	1.288
Upper bound	1.216	1.391	1.565	1.737	1.909

By tuning the frequency for about 10% to change the wavelength, the
gap can also be covered. Therefore, the measurement range of the
single-channel system is any movement amplitude larger than 0.335λ
for a fixed carrier frequency system and larger than $0.214\lambda_{min}$ for a
frequency tunable system, where λ_{min} is the minimum carrier wavelength.

It will be shown in Section 2.3 that when complex signal demodulation is adopted, the limit on even- or odd-order pairs due to residue

phase will be eliminated, resulting in even simpler detection procedure. And the detection range can be further extended.

2.2.3 Pulse Radar

A pulse radar determines target range by measuring the round-trip time of a pulsed microwave signal. Figure 2.6 shows a typical block diagram of a pulse radar.

The transmit/receive (T/R) switch of the pulse radar fulfills two functions: forming the transmit pulse train, and switching the antenna between the transmitter and receiver (duplexing). In the transmit mode, the mixer combines a radio frequency f_0 and an intermediate frequency (IF) f_{IF} to obtain the transmit signal with a frequency of $f_0 + f_{IF}$. After power amplification, pulses of this signal are formed by a pulse generator to give a transmit pulse width τ, with a pulse repetition frequency (PRF) of $f_r = 1/T_r$. The transmit pulse thus consists

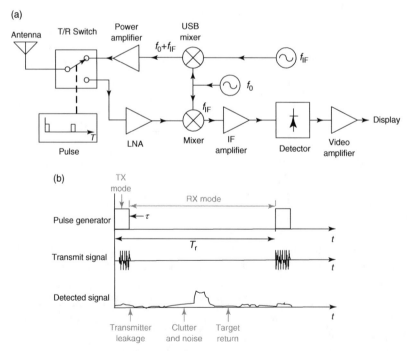

Figure 2.6   (a) Block diagram and (b) operation of a pulse radar. Adapted from Pozar.

of a short burst of a microwave signal at the frequency $f_0 + f_{IF}$. Typical pulse durations range from 100 ms to 50 ns; shorter pulses give better range resolution, but longer pulses result in a better SNR after receiver signal processing. Typical pulse repetition frequencies range from 100 Hz to 100 kHz. Higher PRFs give more returned pulses per unit time, which improves performance, but lower PRFs avoid range ambiguities that can occur when detection distance is too large.

In the receive mode, the returned signal is amplified and mixed with the LO of frequency f_0 to produce the desired IF signal. The same LO is used for both up-conversion and down-conversion to avoid the problems of frequency drift and high phase noise. The IF signal is amplified, detected, and fed into a video amplifier/display. By analyzing the delay between transmitted pulse and received pulse, distance information of the target can be obtained.

2.2.4　FMCW Radar

The pulse radar uses powerful magnetrons to generate microwave signals with very short pulses of applied voltage. In order to overcome the pulse radar's disadvantage of high cost due to the expensive magnetron, the FMCW radar was invented in recent years (Stove, 1992).

In an FMCW radar, a chirp signal is transmitted for a certain duration. While it is transmitting, the echoes are received by the receiver and mixed with the transmit signal, and the result is low-pass filtered to produce a superposition of beat frequencies.

Figure 2.7 shows an example of the simplest format of transmitted signal: a linear chirp. The signal that is reflected from a point target

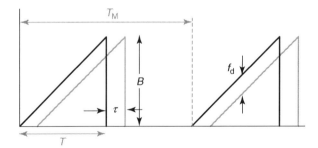

Figure 2.7　Transmit and receive signals of an FMCW radar.

will also be a linear chirp. The round-trip traveling time is τ, the bandwidth of the transmit signal is B, the sweep time is T, and the period is T_M. At any instant in time, the transmitted and received signals are multiplied by a mixer. As multiplying two sinusoidal signals together results in sum and difference terms, after low pass filtering, only the difference term is left at the receiver output. The frequency of this signal is given by f_b, which is the beat frequency. The beat frequency can easily be found as

$$f_b = \frac{B\tau}{T}. \tag{2.33}$$

In a typical FMCW system, the round-trip traveling time is much smaller than the sweep time. As the beat frequency signal is time limited to T seconds, its spectrum will be a sinc function centered at $f = f_b$, and the first zero crossing will occur at $f = 1/T$. When multiple targets exist, the result will be a superposition of many beat frequencies. In the frequency domain, the beat frequencies of two targets can be as close together as $\Delta f_b = 1/T$. Therefore, the difference in traveling time is

$$\Delta\tau = \frac{T\Delta f_b}{B} = \frac{1}{B}. \tag{2.34}$$

Therefore, the minimum separation in time between two targets is inversely proportional to the bandwidth of the FMCW radar.

The range to the target can be calculated as $r = c\tau/2$. The range resolution is given by $\Delta r = c\Delta\tau/2 = c/(2B)$. As the beat frequency is proportional to τ, and τ is proportional to range, knowledge of the beat frequency of any target entails knowledge of the range of that target. With many targets, they can be separated by taking the Fourier transform of the received signal, and the range can be determined through receiver output frequency.

Taking the FMCW radar using linear chirp as an example, the mathematical representation of the transmitted, received, and receiver output can be derived as follows. The transmitted signal can be simplified as

$$T(t) = A\cos\left(2\pi f_0 t + \frac{\pi B}{2T}t^2 + \Phi_1\right), \tag{2.35}$$

where Φ_1 is the residual phase. When the signal is transmitted and reflected by a target, a return signal will be produced. The radar receiver will receive the returned signal as

$$R(t) = A \cdot \Gamma \cdot \cos\left(2\pi f_0 (t - \tau) + \frac{\pi B}{2T}(t - \tau)^2 + \Phi_2\right), \quad (2.36)$$

where Γ is the reflection coefficient magnitude and Φ_2 is the residual phase after reflection. By down-converting the received signal with the same transmitted signal, the radar receiver output is

$$B(t) = \cos(2\pi f_b t + 2\pi f_0 \tau - \pi f_b \tau + \Delta\Phi). \quad (2.37)$$

Variations of modulation are possible (sine, sawtooth, etc.), but the triangle modulation is used in FMCW radars where both range and velocity are desired. With the advent of modern electronics, Digital Signal Processing is used for most detection processing. The beat signals are passed through an A/D converter, and digital processing is performed on the result. Compared with the pulse radar, the FMCW radar can be integrated with solid-state technology, and has the advantages of superior target definition, lower power, and better clutter rejection.

2.2.5 Comparison of Different Detection Mechanisms

Taking the detection of a periodic movement with an arbitrary frequency as an example, the mechanisms of conventional Doppler radar, pulse radar, FMCW radar, and nonlinear Doppler radar are compared. Figure 2.8 shows four types of radar when monitoring a single-tone periodic movement.

The conventional Doppler radar in Fig. 2.8a can measure the frequency or velocity of a moving target based on the Doppler frequency shift. However, unless the instantaneous speed at a high sampling rate is integrated, it does not directly provide position information and thus cannot fully recover the movement pattern.

As shown in Fig. 2.8b, the pulse radar transmits a short pulse of radio signal at each sampling point, and measures the time it takes for the reflection to return. The distance $d_k (k = 1, 2, \ldots, n)$ is one-half of the product of the round-trip time $t_k (k = 1, 2, \ldots, n)$ and the speed of the signal. As radio waves travel at the speed of light, accurate distance measurement requires high performance electronics. The

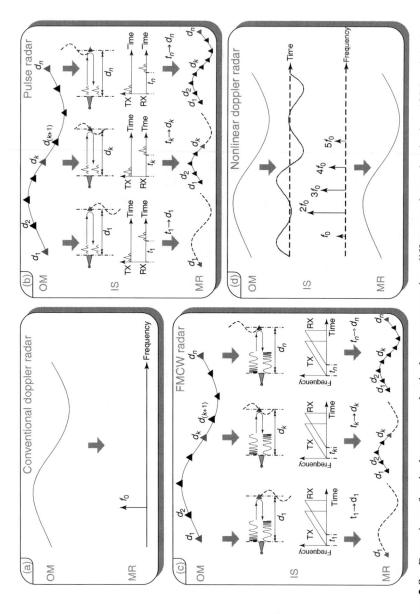

Figure 2.8 Detection of a single-tone mechanical movement using different radar technologies. (OM: original movement; IS: intermediate step; MR: measurement result)

minimum detection range is half the pulse length multiplied by the speed of light. Therefore, a short pulse is needed to detect close targets. On the other hand, longer time between pulses, that is, pulse interval, should be used to maximize the detection range. It is not trivial to tradeoff these two effects and combine both short range and long range in a single radar. As the distance resolution and the characteristics of the received SNR depend heavily on the shape and power of the pulse, the pulse is often modulated to achieve better performance using a pulse compression technique, which increases the cost and sometimes makes the system bulky.

Another alternative is the FMCW radar as shown in Fig. 2.8c. At each sampling point, the FMCW radar transmits a "rising tone" with linearly increasing frequency. The wave propagates from the transmitter, reflects off the subject, and then returns to the receiver. The difference between the received frequency and the transmitted frequency at the same time, coupled with the known rate of frequency increase, allows a time of flight t_k to be calculated, from which distance d_k is calculated.

2.3 KEY THEORY AND TECHNIQUES OF MOTION SENSING RADAR

2.3.1 Null and Optimal Detection Point

The baseband output of a coherent radar detecting a movement $x(t)$ can be generally simplified as $B(t) = \cos[4\pi x(t)/\lambda + \Phi]$, where λ is the carrier wavelength, Φ is the total accumulated residue phase due to the nominal transmission path and phase noise. It is well known that a single-channel radar has the null detection point problem. Intuitively, this is because at certain detection distance, the residue phase Φ renders the cosine function to be close to zero and thus at the radar output the detected signal is close to zero.

In order to fully understand the mechanism of null/optimum detection and obtain a solution, a rigorous analysis can be performed. From the theory of Fourier series, any time-varying periodic displacement $x(t)$ can be viewed as the combination of a series of single-tone vibration. Therefore, for the ease of analysis and without loss of generality, $x(t)$ is assumed to be a single tone, that is, $x(t) = m \sin \omega t$. In this case, the detected phase-modulated signal can be represented as

$$B(t) = \cos\left[\frac{4\pi m \sin(\omega t)}{\lambda} + \Phi\right] = \mathrm{Re}\left(e^{j\left(\frac{4\pi m \sin(\omega t)}{\lambda}\right)} \cdot e^{j\Phi}\right). \quad (2.38)$$

The exponential term can be expanded using Fourier series:

$$e^{j\left(\frac{4\pi m \sin(\omega t)}{\lambda}\right)} = \sum_{n=-\infty}^{\infty} J_n\left(\frac{4\pi m}{\lambda}\right) e^{jn\omega t}, \quad (2.39)$$

where $J_n(x)$ is the nth order Bessel function of the first kind. Therefore, the Fourier series representation of the phase-modulated signal is

$$B(t) = \mathrm{Re}\left(\sum_{n=-\infty}^{\infty} J_n\left(\frac{4\pi m}{\lambda}\right) e^{jn\omega t} \cdot e^{j\Phi}\right)$$

$$= \sum_{n=-\infty}^{\infty} J_n\left(\frac{4\pi m}{\lambda}\right) \cos(n\omega t + \Phi). \quad (2.40)$$

On the basis of the above Fourier expansion, the phase-modulated baseband signal is decomposed into frequency components with n times the basic frequency of the periodic movement. The baseband signal can thus be analyzed in frequency domain. The detection accuracy, however, is affected by the residual phase Φ. To address this issue, the $n < 0$ and $n > 0$ terms in Equation 2.40 should be combined. The negative-order and positive-order Bessel functions are related by

$$J_n(x) = \begin{cases} J_{-n}(x) & \text{for even } n \\ -J_{-n}(x) & \text{for odd } n \end{cases}. \quad (2.41)$$

Therefore, the baseband output can be reduced to

$$\widehat{B}(t) = \sum_{k=0}^{\infty} J_{2k+1}\left(\frac{4\pi m}{\lambda}\right) [\cos(k\omega t + \Phi) - \cos(\Phi - k\omega t)]$$

$$+ \sum_{k=0}^{\infty} J_{2k}\left(\frac{4\pi m}{\lambda}\right) [\cos(k\omega t + \Phi) + \cos(\Phi - k\omega t)]$$

$$= -2 \cdot \sum_{k=1}^{\infty} J_{2k}\left(\frac{4\pi m}{\lambda}\right) \cdot \cos 2k\omega t \cdot \cos \Phi$$

$$- 2 \cdot \sum_{k=0}^{\infty} J_{2k+1}\left(\frac{4\pi m}{\lambda}\right) \cdot \sin(2k+1)\omega t \cdot \sin \Phi. \quad (2.42)$$

Note that the term $J_0(4\pi m/\lambda) \cos \Phi$ is neglected as this is the DC term and has nothing to do with successful detection. The last two terms of Equation 2.42 correspond to the odd-order and even-order harmonics in baseband spectra, respectively. On the basis of this equation, successful detection of the periodic movement, which is related to the fundamental frequency ($k = 0$ in the last term of the above detection), is also dependent on the residual phase Φ. Taking two extreme cases as an example: When Φ is equal to odd multiples of $90°$, the even-order frequencies vanish, thus the desired fundamental frequency is emphasized while even-order harmonics are minimized. When Φ is equal to even multiples of $90°$, the odd-order frequencies including the desired fundamental frequency vanish.

The residual phase Φ is contributed by two factors: the constant phase shift due to the distance to the target d_0 and reflections at the target surface, and the total phase noise. Because of the range correlation effect, the phase noise plays a minor role and Φ is, to a large extent, determined by the distance to the target d_0

$$\phi = \theta + \Delta\phi(t) = \frac{4\pi d_0}{\lambda} + \tilde{\phi}, \quad (2.43)$$

where $\tilde{\phi}$ can be treated as unchanged during experiment. Therefore, a series of optimum points and null points for accurate detection exist along the path from the radar to the target.

2.3.2 Complex Signal Demodulation

The quadrature radar has I/Q channel outputs that are out of phase, that is, $I(t) = \cos[4\pi x(t)/\lambda + \Phi]$ and $Q(t) = \sin[4\pi x(t)/\lambda + \Phi]$. The complex signal demodulation can eliminate the optimum/null detection point problem by combining the I and Q signals in baseband. As shown in Fig. 2.9, the complex signal can be reconstructed

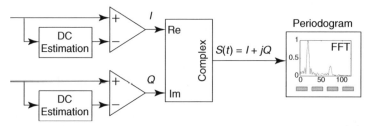

Figure 2.9 Block diagram of complex signal demodulation (Li and Lin 2008a).

in software in real time as

$$S(t) = I(t) + j \cdot Q(t) = \exp\left\{ j \left[\frac{4\pi x\,(t)}{\lambda} + \Phi \right] \right\}$$

$$= \sum_{k=-\infty}^{\infty} J_k \left(\frac{4\pi m}{\lambda} \right) \cdot e^{jk\omega t} \cdot e^{j\Phi}. \tag{2.44}$$

As $e^{j\Phi}$ has a constant envelope of unity, the effect of Φ on signal amplitude can be eliminated. Applying the complex Fourier transform to the signal $S(t)$ for spectral analysis, the residual phase Φ will not affect the relative strength between the odd-order and the even-order frequency components. The desired fundamental frequency component will always be present in the spectrum.

Meanwhile, even though DC offsets exist in the I/Q channels and may lead to the error of measured DC component, they only affect the DC term of $S(t)$. Therefore, the existence of DC offset does not affect obtaining the information of the desired frequency tones. In practice, the residual baseband DC level that is the sum of the motion-related movement and the DC offset, can be easily extracted as the average of signals in every time-domain sliding window and thus be safely removed. As a result, the complex signal demodulation greatly simplifies the demodulation procedure and is immune from DC offset.

2.3.3 Arctangent Demodulation

Another way to eliminate the optimum/null detection point problem in the quadrature demodulation system is to use arctangent demodulation (Park et al., 2007) by calculating the total Doppler phase shift. Its

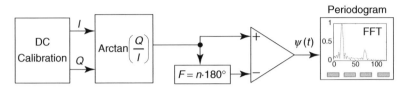

Figure 2.10   Block diagram of arctangent demodulation.

block diagram is shown in Fig. 2.10. Taking into account the phase discontinuity when the signal trajectory crosses the boundary of two adjacent quadrants, the arctangent demodulation calculates the total angular information $\psi(t)$ as

$$\psi(t) = \arctan\frac{Q(t)}{I(t)} + F = \frac{4\pi x(t)}{\lambda} + \Phi, \qquad (2.45)$$

where F is a multiple of $180°$ for the purpose of eliminating the discontinuity when $\psi(t)$ crosses the boundary of two adjacent quadrants in the constellation graph.

Because $\psi(t)$ has a linear relationship to the movement $x(t)$, the motion information can be obtained with the nonlinear phase modulation effect eliminated. The advantage is the ability to eliminate the harmonic and intermodulation interference. However, this technique requires accurate calibration of the DC offset in order to properly reconstruct the angular information. The difficulty of accurate DC offset calibration encountered in Doppler radar motion sensing is that the DC offset is not only produced by electronic circuits, but also by the unmodulated reflected signal, that is, signal reflected from stationary objects and other stationary parts of the target rather than the moving parts. Therefore, the DC offset changes as the environment changes and needs to be recalibrated once the change occurs.

On the other hand, the presence of baseband DC offsets in I/Q channels leads to a shifted trajectory in the constellation graph. Although, the angular information $\psi(t)$ will be changed significantly when the trajectory is shifted, the angular movement is still periodic. This implies that when analyzing the spectrum of $\psi(t)$ in the presence of a DC offset, the desired frequency components still exist. The difference observed in the spectrum is a changed harmonic level. Therefore, if the DC offset can be properly estimated in software, it is still possible to extract the desired motion signal without a hardware-based DC offset calibration. As will be shown later, a trajectory-fitting procedure can be adopted for DC offset estimation in the baseband.

2.3.4 Double-Sideband Transmission

For single-carrier transmission, the distance between null points is decided by

$$\frac{4\pi\Delta d}{\lambda} = \pi \Rightarrow \Delta d = \frac{\lambda}{4}. \tag{2.46}$$

If two carriers are used for transmission with a slightly difference in Δd because of wavelength difference, the occurrence of global null point, where both carriers are at their respective null points, can be largely reduced.

The block diagram of double-sideband radar architecture is shown in Fig. 2.11. Two LOs, LO1 and LO2, are used to generate an IF of f_1 and an RF of f_2 respectively. The up-converter mixes the two signals together to generate the upper sideband signal $f_U = f_2 + f_1$ and the lower sideband signal $f_L = f_2 - f_1$.

If there are two frequency components f_L and f_U in the transmitted signal $T(t)$, the received signal $R(t)$ will correspond to these two frequency components f_L and f_U as well. Let $B_L(t)$ and $B_U(t)$ represent the baseband signals corresponding to f_L and f_U, respectively. In this case, the total baseband output will be

$$B(t) = B_L(t) + B_U(t), \tag{2.47}$$

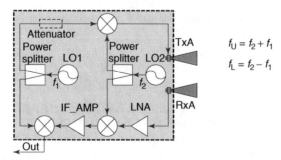

Figure 2.11 Block diagram of double-sideband radar architecture.

where

$$B_{\rm L}(t) = \cos\left[\theta_{\rm L} + \frac{4\pi x\,(t)}{\lambda_{\rm L}} + \Delta\Phi_{\rm L}(t)\right], \tag{2.48}$$

$$B_{\rm U}(t) = \cos\left[\theta_{\rm U} + \frac{4\pi x\,(t)}{\lambda_{\rm U}} + \Delta\Phi_{\rm U}(t)\right], \tag{2.49}$$

and

$$\theta_{\rm L} = \frac{4\pi d_0}{\lambda_{\rm L}} + \theta_{0{\rm L}}, \; \theta_{\rm U} = \frac{4\pi d_0}{\lambda_{\rm U}} + \theta_{0{\rm U}}, \tag{2.50}$$

where $\lambda_{\rm L}$ and $\lambda_{\rm U}$ are wavelengths of lower sideband and upper sideband, which equal to $c/f_{\rm L}$ and $c/f_{\rm U}$, respectively. $\theta_{\rm L}$ and $\theta_{\rm U}$ are fixed phase shifts of the lower sideband signal and the upper sideband signal, respectively.

From the above discussions, either $B_{\rm L}(t)$ or $B_{\rm U}(t)$ has severe null detection point problem and cannot give a reliable detection at high frequency. However, when $B_{\rm L}(t)$ and $B_{\rm U}(t)$ simultaneously exist, $B(t)$ is the superposition of $B_{\rm L}(t)$ and $B_{\rm U}(t)$. $B_{\rm L}(t)$ and $B_{\rm U}(t)$ are similar but have different spatial variation between them. If this difference is arranged properly, the baseband output $B(t)$ should not have the severe null-point problem that either $B_{\rm L}(t)$ or $B_{\rm U}(t)$ alone has. Figure 2.12 shows the distribution of null points and optimum points for each carrier frequency of double-sideband transmission. If the LO1 frequency f_1 is arranged properly, the null points from lower sideband and optimum points from upper sideband, or vice versa, can overlap

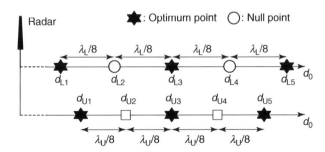

Figure 2.12 Optimum points and null points distribute along the path away from the radar for double-sideband transmission. The subscript L represents lower sideband, and U represents upper sideband (Xiao et al., 2006).

each other. Good detection accuracy is therefore achieved over a wide distance range.

As the residual phase noises $\Delta\Phi_L(t)$ and $\Delta\Phi_U(t)$ are much smaller compared with θ and the phase modulation $4\pi x(t)/\lambda$, because of the effect of range correlation, they will be ignored in the following analysis. In addition, phase modulations $4\pi x(t)/\lambda_L$ and $4\pi x(t)/\lambda_U$ have nearly the same amplitudes because λ_L is very close to λ_U.

When θ_L and θ_U are separated by an even multiple of π, $B_L(t)$ and $B_U(t)$ become the same. Therefore, $B(t)$ will give the same optimum points and null points at the same places as those given by either $B_L(t)$ or $B_U(t)$ alone, and has the same problem of closely spaced null points that degrade the detection accuracy and reliability.

When θ_L and θ_U are separated by an odd multiple of π, $B_L(t)$ and $B_U(t)$ have equal same amplitude but opposite signs, hence canceling each other. Therefore, $B(t)$ becomes very small and hard to be detected.

As a result, when the phase difference between θ_L and θ_U is an integer multiple of π, a new null-point condition occurs in the measurement. If the null point of the single-sideband transmission is defined as the local null point, then this new null-point condition is defined as the global null point. At this global null point, the detection accuracy is the lowest.

Let

$$\theta_U - \theta_L = \frac{4\pi d_0}{\lambda_U} - \frac{4\pi d_0}{\lambda_L} + \Delta\theta_0 = k\pi, \quad k = 0, \pm1, \pm2, \ldots \tag{2.51}$$

where

$$\Delta\theta_0 = \theta_{0U} - \theta_{0L}. \tag{2.52}$$

Substituting $\lambda_L = c/f_L$, $\lambda_U = c/f_U$, then

$$f_U - f_L = \frac{c}{4\pi d_0}(k\pi - \Delta\theta_0), \quad k = 0, \pm1, \pm2, \ldots \tag{2.53}$$

Substituting $f_U = f_2 + f_1$, and $f_L = f_2 - f_1$, then

$$f_1 = \frac{k}{d_0} \cdot 37.5 \text{ MHz} - \frac{c}{8\pi d_0} \cdot \Delta\theta_0, \quad k = 0, \pm1, \pm2, \ldots \tag{2.54}$$

where d_0 is the distance in meters.

When θ_L and θ_U are separated by an odd multiple of $\pi/2$, $B_L(t)$, and $B_U(t)$ are effectively in quadrature. At least one of $B_L(t)$ and $B_U(t)$ is not at the null point. The one that is not at the null point will be dominant in the final output $B(t)$. Therefore, in this case, the overall detection accuracy will be high. This point is defined as the global optimum point. Let

$$\theta_U - \theta_L = \frac{4\pi d_0}{\lambda_U} - \frac{4\pi d_0}{\lambda_L} + \Delta\theta_0 = k\pi + \frac{\pi}{2}, \quad k = 0, \pm 1, \pm 2, \ldots$$

(2.55)

Repeat the same process, it can be shown that

$$f_1 = \frac{2k+1}{d_0} \cdot 18.75 \text{ MHz} - \frac{c}{8\pi d_0} \cdot \Delta\theta_0, \quad k = 0, \pm 1, \pm 2, \ldots$$

(2.56)

where d_0 is the distance in meter.

The frequency difference between f_U and f_L is $2f_1$. Therefore, the selection of f_1 will determine if θ_L and θ_U are separated by $k\pi$ or $k\pi + \pi/2$, and thus whether the subject's position is at a null point or an optimum point.

When θ_L and θ_U are separated by an arbitrary angle other than $k\pi$ and $k\pi + \pi/2$, or f_1 is between the above two cases, the detection accuracy will be between the above two cases.

The above analysis shows that when the position of the subject is fixed, this position can be set to a global optimum point or a global null point by properly choosing the frequency for f_1. For example, if at a given f_1, the subject position at $d_0 = 1$ m happens to be a null point, this null point can be changed to an optimum point if f_1 is tuned to $f_1 \pm (2k+1) \times 18.75$ MHz according to (2.54) and (2.56). This means that an accurate detection can always be made at an optimum point by adjusting f_1 without moving the subject's position.

When the frequency f_1 is fixed, the distribution of the global null points and optimum points for double-sideband transmission is different from single-sideband case due to the superposition of two baseband signals. Rewriting equations (2.54) and (2.56) as

$$d_0 = \frac{k}{8}\lambda_1 - \frac{\lambda_1}{8\pi} \cdot \Delta\theta_0, \quad k = 0, \pm 1, \pm 2, \ldots$$ (2.57)

and

$$d_0 = \frac{(2k + 1)}{16} \cdot \lambda_1 - \frac{\lambda_1}{8\pi} \cdot \Delta\theta_0, \quad k = 0, \pm 1, \pm 2, \ldots \quad (2.58)$$

for the conditions of global null point and global optimum point, respectively. From the above equations, the global null points are encountered every $\lambda_1/8$, so are the global optimum points. Furthermore, adjacent global null point and global optimum point are separated by $\lambda_1/16$. The distribution of the global null points and the global optimum points for double-sideband transmission is shown in Fig. 2.13. As the LO1 frequency f_1 is much smaller than the LO2 frequency f_2, the distance between adjacent global null point and global optimum point is much larger than the single-sideband case. Taking a Ka-band double-sideband radar (Xiao et al., 2006) as an example for $f_1 = 500$ MHz, which is much smaller than any Ka-band frequency, the null point occurs every 75 mm. This is much larger than the null point separation of 2.5 mm for a single-carrier 30 GHz system. Therefore, by using double-sideband transmission, it is possible to obtain reliable detection accuracy and avoid the null point problem by adjusting the position of the radar.

From the above equations, it seems that the lower the f_1 is, the further the null points are separated and thus the null-point problem would be solved with very low f_1. When f_1 is too small, however, the null points will be dominated by the local null points over a wide range in distance. Figure 2.14 shows the distribution of the local null points and the global null points for $f_2 = 27.1$ GHz and $f_1 = 500$ MHz, 50 MHz, and 5 MHz, respectively. The y-axis indicates the normalized amplitude of the signals. When the signal hits the valley, the amplitude

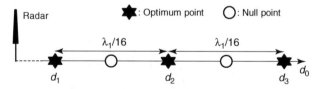

Figure 2.13 Global optimum points and null points for double-sideband transmission along the path away from the radar. d_1 and d_3 are optimum points. d_2 is null point. The adjacent global null point and optimum point are separated by $\lambda_1/16$ (Xiao et al., 2006).

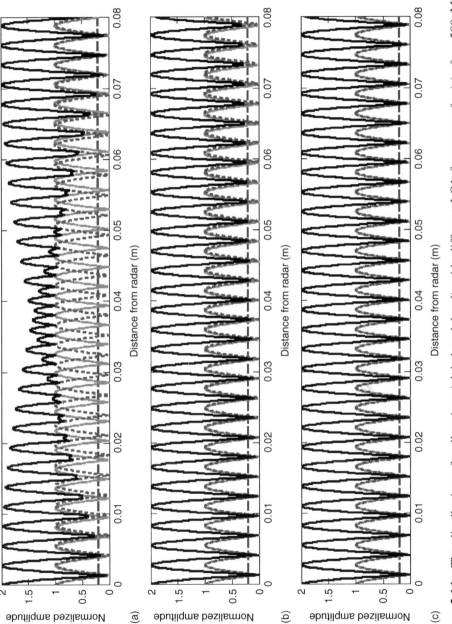

Figure 2.14 The distribution of null points (global and local) with different LO1 frequency f_1. (a) $f_1 = 500$ MHz, (b) $f_1 = 50$ MHz, and (c) $f_1 = 5$ MHz. The dashed line indicates 20% of peak amplitude.

41

is the smallest, thus the detection accuracy is the lowest. The light solid and the light dotted lines show the distribution of the local null points and the local optimum points for baseband signals $B_L(t)$ and $B_U(t)$, respectively. The amplitude of $B_L(t)$ and $B_U(t)$ may have a little difference because of frequency response flatness in transceiver, but here the same amplitude is assumed for the convenience of analysis.

As shown in Fig. 2.14, the separation of the nearest local null points (valley) is about 2.5 mm. The thick solid lines show the distribution of the global null points and the global optimum points for $B(t)$. When $f_1 = 500$ MHz, the separation of the global null points is 75 mm, which is shown in Fig. 2.14a. However, for $f_1 = 5$ MHz, the separation of the global null points is 7.5 m. As shown in Fig. 2.14c, within a 0.1 m range, $B(t)$ has the same null points and optimum points as those of $B_L(t)$ or $B_U(t)$, which was qualitatively defined as a global null point in previous analysis. Quantitatively, if the signal valley amplitude for $B(t)$ falls under 20% of peak amplitude of either $B_L(t)$ or $B_U(t)$, then we define this condition as the global null-point region. By this definition, $B(t)$ will stay in a global null-point region for about 1-m long for $f_1 = 5$ MHz, 0.1 m for 50 MHz, and 0.01 m for 500 MHz, respectively.

To overcome the null-point problem in the measurement, and to obtain high detection accuracy, it is better to make the measurement at or near the optimum point by either moving the radar position or changing the f_1 frequency. For f_1 as low as 5 MHz, sometimes it is hard to move the system as much as 3 m in distance for it to reach a nearest optimum point. Therefore, the best way is to adjust the LO1 frequency, f_1.

For short-range motion sensing applications, if a null point occurs at $d_0 = 2.5$ m, in order to switch this null point to an optimum point, the f_1 frequency will need to be changed at least 7.5 MHz according to Equation 2.56. However, if a null point occurs at $d_0 = 0.1$ m, the smallest tuning step will be 187.5 MHz that is quite a large tuning range for LO1. Therefore, the selection of f_1 frequency and the voltage-controlled oscillator (VCO) tuning range need to be considered together when the null point appears at a distance close to radar.

In summary, the global null points for double-sideband transmission are encountered every $\lambda_1/8$, where λ_1 is the wavelength corresponding to the IF LO with a frequency much lower than RF carrier frequency.

The distance between null points is made much larger by double-sideband transmission. Furthermore, double-sideband transmission can always make an accurate detection by slightly tuning the IF LO to change an arbitrary position into an optimum point.

2.3.5 Optimal Carrier Frequency

For radar based on Doppler nonlinear phase modulation mechanism, it is important to select an optimal carrier frequency for motion sensing because of the nonlinear transfer function. Without loss of generality, the baseband output spectrum of a quadrature radar can be expressed as the sum of frequency tones of $J_k(4\pi m/\lambda) \cdot \exp(jk\omega t) \cdot \exp(j\varphi)$, where k is any integral number and represents the number of harmonics. The amplitude of each frequency tone is thus decided by the coefficient $J_k(4\pi m/\lambda)$, which is the Bessel function of the first kind. Figure 2.15 shows the plot of Bessel functions. It can be seen from the plot that there is an optimal frequency that can maximize each of the frequency tones in baseband output.

For the following, an example of Doppler radar noncontact detection of respiratory and heartbeat motion will be given to illustrate the importance of properly choosing the optimal carrier frequency. In noncontact vital sign detection, the complex signal demodulated radar

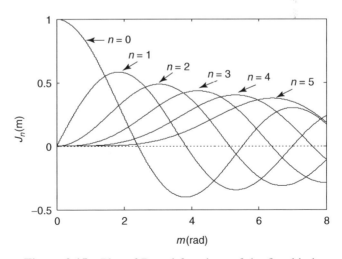

Figure 2.15 Plot of Bessel functions of the first kind.

baseband output will be

$$B(t) = \exp\left[j\left(\frac{4\pi x_h(t)}{\lambda} + \frac{4\pi x_r(t)}{\lambda} + \phi\right)\right], \qquad (2.59)$$

where $x_r(t) = m_r \cdot \sin\omega_r t$ and $x_h(t) = m_h \cdot \sin\omega_h t$ represent body movements due to respiration and heartbeat, respectively. The corresponding Fourier series representation of the spectrum is

$$B(t) = \sum_{k=-\infty}^{\infty} J_k\left(\frac{4\pi m_r}{\lambda}\right) e^{jk\omega_r t} \sum_{l=-\infty}^{\infty} J_l\left(\frac{4\pi m_h}{\lambda}\right) e^{jl\omega_h t} \cdot e^{\phi}$$

$$= \sum_{k=-\infty}^{\infty} \sum_{l=-\infty}^{\infty} J_l\left(\frac{4\pi m_h}{\lambda}\right) J_k\left(\frac{4\pi m_r}{\lambda}\right) e^{j(k\omega_r t + l\omega_h t + \phi)}. \qquad (2.60)$$

The above result shows that the nonlinear property of cosine transfer function not only causes the undesired effect of harmonics interference, but also causes intermodulation between respiration signal and heartbeat signal. Therefore, the detected strength of a desired signal (respiration or heartbeat) is determined by both the signal itself and the other signal (heartbeat or respiration). For example, the detected heartbeat signal is determined by the ($l = \pm 1$, $k = 0$) terms, and its amplitude $J_{\pm 1}(4\pi m_h/\lambda) \cdot J_{\pm 0}(4\pi m_r/\lambda)$ is dependent on both m_r and m_h.

To verify the effects of intermodulation and harmonics, the mechanism of Doppler radar vital sign detection is also modeled in Agilent Advanced Design System (ADS) as shown in Fig. 2.16. The phase shift Φ is realized by a phase shifter, labeled as PS; and the body movement caused by respiration and heartbeat is modeled by two cascaded sinusoidal signal sources SRC1 and SRC2.

Typically, relaxed human beings have an effective m_h on the order of 0.01 mm, and an effective m_r varying from around 0.1 mm to several millimeters. As an illustration of the nonlinear property in Doppler radar vital sign detection, Fig. 2.17 represents the simulated baseband spectrum when $m_r = 0.8$ mm and $m_h = 0.08$ mm.

As shown in Fig. 2.17, besides the desired breathing signal (B_1) and heartbeat signal (H_1), harmonics of respiration (B_2 and B_3) as well as

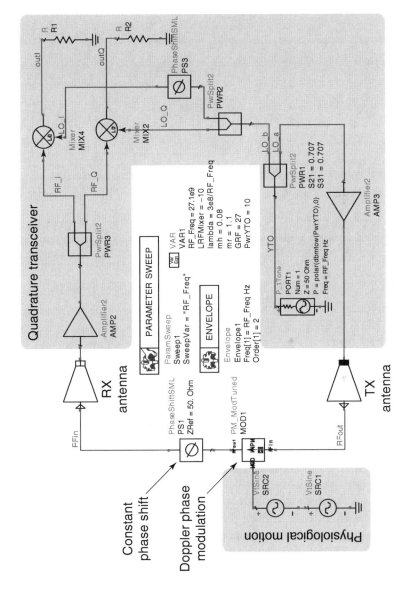

Figure 2.16 Modeling of Doppler radar vital sign detection with quadrature transceiver; PS produces the total phase shift Φ (Li and Lin, 2007b).

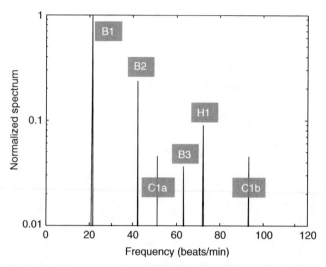

Figure 2.17   Simulated spectrum of baseband signal; B_1, B_2, and B_3: breathing fundamental, the second harmonic, and the third harmonic; H_1: heartbeat fundamental; C_{1a} and C_{1b}: lower sideband and upper sideband caused by the intermodulation of B_1 and H_1.

the intermodulation terms (C_{1a} and C_{1b}) were observed. As the respiration signal (B_1) is usually the lowest frequency component in the baseband spectrum and respiration usually causes a larger chest wall movement than heartbeat does, it is easy to extract respiration signal while accurate detection of heart rate presents the main challenge for vital sign detection. The following simulations and discussions will be focused on the detection of heartbeat signal H_1.

The harmonics of respiration (B_3, B_4, etc.) have frequencies varying with the respiration rate, and thus may be very close to heartbeat frequency and significantly deteriorate the detection accuracy. From this point of view, the absolute strength of detected heartbeat signal, and the strength of detected heartbeat signal compared with harmonics of respiration signal are important factors influencing the performance of the vital sign detector. These factors are investigated as follows.

2.3.5.1 Absolute Detected Heartbeat Strength.   The theoretical (solid line) and the ADS-simulated (markers) detected heartbeat strength as a function of the carrier frequency is shown in Fig. 2.18.

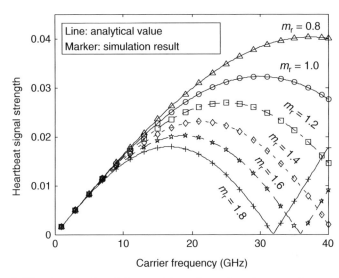

Figure 2.18 Amplitude of detected heartbeat signal versus the carrier frequency; the amplitude of respiration-induced body movement m_r ranges from 0.8 to 1.8 mm; the amplitude of heartbeat-induced body movement was assumed to be $m_h = 0.08$ mm.

The amplitude of heartbeat-induced body movement was assumed to be $m_h = 0.08$ mm, which is a typical value of ordinary people. And the amplitude of respiration-induced body movement was considered ranging from 0.8 to 1.8 mm, assuming the most challenging case when detecting from the front of the human body.

As shown in the result, the amplitude of detected heartbeat signal increases with the carrier frequency when the carrier frequency is lower than 17 GHz. However, when the carrier frequency becomes large enough, the signal amplitude begins to decrease, especially for large m_r. According to the Equation 2.60, this can be explained as follows: as the carrier frequency increases, the decreasing speed of $J_0(4\pi m_r/\lambda)$ begins to exceed the increasing speed of $J_{\pm 1}(4\pi m_h/\lambda)$.

Moreover, the detected heartbeat signal may even disappear at certain frequencies for a fixed m_r producing a detection absolute-null-point. This detection absolute-null-point is caused by the null point of Bessel function. It should be noted that the absolute-null-point cannot be eliminated by adjusting the phase shift Φ, and is different from the null points discussed in the previous sections, where those null points for certain phase shift Φ can be eliminated by either double-sideband transmission or quadrature demodulation.

Therefore, for the sake of maximizing the detected heartbeat strength, there is an optimum carrier frequency for a fixed value of m_r For instance, the optimum carrier frequency when $m_r = 1.2$ mm is approximately 27 GHz.

2.3.5.2 Detected Heartbeat Compared with Harmonics.

Figure 2.19 shows the theoretical (solid line) and simulated (markers)-detected heartbeat signal strength compared with the third order breathing harmonic. It is shown that as the carrier frequency increases, the relative strength decreases until it reaches a null point, which is caused by the absolute-null-point of detection.

To reduce the harmonics interference, the carrier frequency should be confined to a range defined by a critical value. Assuming that this critical frequency is set where the heartbeat signal has equal strength as the third order harmonic, a dashed line was drawn in Fig. 2.19.

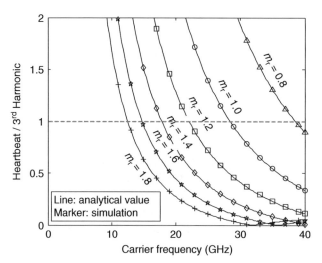

Figure 2.19 Relative strength of detected heartbeat signal compared with the third order harmonic as the carrier frequency changes; the amplitude of respiration-induced body movement m_r ranges from 0.8 to 1.8 mm; the amplitude of heartbeat-induced body movement was assumed to be $m_h = 0.08$ mm.

The intersections indicate the critical frequencies for different m_r values. For example, when m_r is equal to 1.0 mm, the maximum carrier frequency is set around 29 GHz.

In summary, in contrast to the common sense that detection accuracy can always be increased by increasing the carrier frequency, there is an optimum choice of carrier frequency. At the optimum carrier frequency, the desired signal component can be maximized in the premise that the harmonics and intermodulation interference are not so large as to affect the detection accuracy. Taking the noncontact physiological motion sensing application as an example, the carrier frequency can be increased up to the lower region of the Ka-band to improve detection accuracy.

2.3.6 Sensitivity: Gain and Noise Budget

In communication systems, the receiver sensitivity is usually given in logarithm scale by

$$\text{Sensitivity} = kTB\,(\text{dBm}) + F\,(\text{dB}) + \text{SNR}(\text{dB}), \qquad (2.61)$$

where kT is the input thermal noise floor per unit bandwidth, B is the bandwidth of the receiver, F is the noise figure of the receiver, and SNR is the signal-to-noise-ratio required for baseband signal processing. When the Equation 2.61 is applied to radar motion sensing, it results in an overly pessimistic estimation of the system performance. In such applications, the radar normally detects a low frequency signal caused by mechanical movement. For example, the physiological sensor detects the heartbeat signal around 1 Hz. The noise figure of a complementary metal-oxide semiconductor (CMOS) receiver chip could be as high as 30 dB due to flicker noise contribution at such a low frequency. If the receiver has a baseband bandwidth of 1 MHz and the SNR requirement is set to 0 dB for analog demodulation, the sensitivity is $-174\,\text{dBm/Hz} + 10 \times \log_{10}(1\,\text{MHz}) + 30\,\text{dB} = -84\,\text{dBm}$, which corresponds to a very limited detection range as the useful signal is weak when reflected from the target.

In reality, the baseband output noise can be represented as in Fig. 2.20. At very low frequency, flicker noise in baseband circuits dominates. It rolls-off at a slope of -10 dB/decade until intersecting the flicker noise corner, after which white noise becomes the main noise

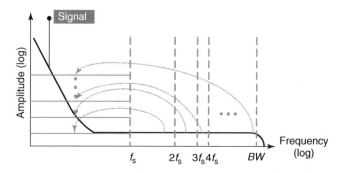

Figure 2.20   Noise figure as a function of baseband output frequency.

source. Simply measuring the noise figure around the signal frequency is not applicable over the entire receiving baseband. Instead, the amplified flicker noise and white noise should be considered separately for accurate estimation of receiver sensitivity. Another important factor for the system sensitivity is the sampling rate f_s of the baseband ADC. Because the physical motion to be detected has a very low frequency, usually a low sampling rate was used to reduce the cost of the ADC and the digital signal processor. As a result, the amplified white noise is folded into the frequency band from DC to f_s due to aliasing, as indicated in Fig. 2.20.

For real-time signal processing of motion detection, a Fourier transform is usually applied to the baseband signal, so that the signal frequency can be identified from the resultant periodogram. In the periodogram, the noise power is distributed into a resolution bandwidth (RBW) determined by the time-domain window size of the Fourier transform. The narrower the RBW, the less noise energy is contained in a single point of the periodogram, thus the higher the output SNR. Taking into account the effect of noise aliasing and RBW, the receiver output noise level on the periodogram can be quantified as

$$\text{Flicker noise} = kT \cdot \text{RBW} \cdot F_{\text{flicker}}(f) \cdot G, \qquad (2.62)$$

$$\text{White noise} = kT \cdot \text{RBW} \cdot F_{\text{white}} \cdot \left(\frac{B}{f_s}\right) \cdot G, \qquad (2.63)$$

where G is the gain of receiver chain, F_{white} is the noise figure at frequencies above the flicker noise corner frequency, $F_{\text{flicker}}(f)$ is the noise figure contributed by flicker noise, B is the equivalent noise bandwidth of the receiver, RBW $= 1/\text{TW}$ is the RBW of the periodogram, which is decided by the Fourier transform window size TW. For a single-pole system, B is approximated as $\pi/2$ times the 3-dB bandwidth. In order to reduce the noise level on the periodogram, a large window size TW is desired for the Fourier transform. However, in motion sensing, the radar needs to track real-time variation of the target position. Also, there are limited memory size and digital signal processing speed in portable applications. Therefore, practical values of TW have a limited size. For example, a window size of 10–12.5 s has been used in typical vital sign monitoring applications, resulting in an RBW of around 0.1 Hz.

On the basis of the noise level described above, the sensitivity of a motion detection radar can now be estimated as (in linear scale)

$$\text{Sensitivity} = \left\{ kT \cdot \text{RBW} \cdot F_{\text{white}} \cdot \left(\frac{B}{f_s} \right) + kT \cdot \text{RBW} \cdot F_{\text{flicker}}(f) \right.$$

$$\left. + \text{PN}_{\text{LO}} + e_{\text{ADC}} \right\} \cdot \text{SNR}, \tag{2.64}$$

where PN_{LO} is the phase noise contribution from the LO, e_{ADC} is the ADC quantization error referred to the receiver input, and SNR is the signal-to-noise ratio required on the periodogram for reliable detection. It should be noted that the LO phase noise is largely reduced because of the radar range correlation effect. A well-designed free running VCO has negligible effect on the receiver sensitivity. Therefore, PN_{LO} is neglected in the following discussion. It has also been shown by Yan et al. that the LO I/Q mismatch affects the distribution of null detection point and reduces the accuracy when detecting the movement amplitude using the nonlinear phase modulation, but it does not affect the sensitivity at the optimal detection point (Yan et al., 2010).

Taking the noncontact vital sign detection as an example, normally a 10 to 20 dB SNR is good enough for reliable detection. Assuming $F_{\text{flicker}} = 40$ dB at 1 Hz, $F_{\text{white}} = 8$ dB, $B = 1$ MHz, RBW $= 0.1$ Hz, and $f_s = 20$ Hz, the input-referred white noise is -174 dBm/Hz $+ 10 \times \log_{10}(0.1 \text{ Hz}) + 10 \times \log_{10}(1 \text{ MHz}/20 \text{ Hz}) + 8$ dB$+ = -129$ dBm, while the input-referred flicker noise is -174 dBm/Hz $+ 10 \times \log_{10}(0.1 \text{ Hz}) + 40$ dB$+ = -144$ dBm. Neglecting ADC quantization error and LO residual phase noise, the noise produced in the receiver chain is much less than the input-referred noise of −84 dBm estimated based on (2.61), even though this example used worse flicker noise than the example based on (2.61). Therefore, it is shown that the real sensitivity predicted by (2.64) could be much better than that estimated by (2.61). It should be noted that the relative strength of flicker noise and white noise is controlled by several parameters such as f_s and B. If a high sampling rate of 10 kHz was used, the input-referred white noise will be reduced to −156 dBm, making the flicker noise dominant again in the receiver chain.

For portable applications, f_s is decided by the availability of low power, low cost ADC. RBW is determined by the specific application, the memory size, and the digital signal processing speed. Therefore, on the receiver design, the focus is to reduce the flicker noise $F_{\text{flicker}}(f)$ and control the strength and bandwidth of white noise. Although, the receiver chip noise figure could be very high around the signal of interest due to large flicker noise, it is possible to design a high sensitivity receiver by reducing the sum of noise contribution from both flicker noise and white noise. To realize this, system consideration of link budgets are discussed as follows.

Assume a direct-conversion motion sensing radar has the following stages cascaded from RF input to baseband output: low noise amplifier (LNA), preamplifier (Preamp), mixer, and baseband variable gain amplifier (VGA). The noise figure of the receiver chain can be expressed as

$$F = F_{\text{LNA}} + \frac{F_{\text{Preamp}} - 1}{G_{\text{LNA}}} + \frac{F_{\text{mixer}} - 1}{G_{\text{LNA}}G_{\text{Preamp}}} + \frac{F_{\text{VGA}} - 1}{G_{\text{LNA}}G_{\text{Preamp}}G_{\text{mixer}}}. \tag{2.65}$$

If the baseband VGA has a noise figure of 60 dB at 1 Hz baseband frequency, while the preceding LNA and preamplifier has a total gain of 40 dB, the receiver noise figure contributed by VGA will be approximately 20 dB at 1 Hz output frequency. In this case, the receiver noise figure may no longer be dominated by the LNA. Therefore, different from traditional communication systems, large gains in the radio frequency blocks (LNA + Preamplifier) and low baseband flicker noise (mixer + VGA) are emphasized in circuit design of Doppler vital sign detection radar.

3

HARDWARE DEVELOPMENT OF MICROWAVE MOTION SENSORS

3.1 RADAR TRANSCEIVER

Microwave transceiver is the key building block of a noncontact motion sensing radar. The complexity of transceiver architecture for motion sensing radar varies a lot, depending on the requirement of detection range and accuracy. For short-range detection, the transceiver architecture can be fairly simple, which makes it easier for micro-radar system-in-package or system-on-chip integration. Several radar transceiver developments will be introduced in this section as follows.

3.1.1 Bench-Top Radar Systems

Some of the milestone motion-sensing bench-top radar systems were built at the Michigan State University by Chen et al. for microwave life-detection applications. The first emerging problem solved was the choice of RF. In the 1980s, a microwave life-detection system, which

Microwave Noncontact Motion Sensing and Analysis, First Edition.
Changzhi Li and Jenshan Lin.
© 2014 John Wiley & Sons, Inc. Published 2014 by John Wiley & Sons, Inc.

operates at the X-band (10 GHz) for sensing the physiological status of soldiers lying on the ground of a battlefield, was reported (Chen et al., 1986; Chuang et al., 1991). It turned out, though, that such an X-band microwave beam cannot penetrate earthquake rubble or collapsed building debris deep enough to locate buried human victims. For this reason, two other systems, one operating at 450 MHz (Chen et al., 1994) and the other at 1150 MHz (Chen et al., 1996), have been implemented. On the basis of a series of experiments (Chen et al., 2000a), it was found that an electromagnetic (EM) wave of 1150 MHz can penetrate earthquake rubble (layers of reinforced concrete slabs) with metallic wire mesh easier than an EM of 450 MHz. However, an EM wave of 450 MHz may penetrate deeper into rubble without metallic wire mesh than an EM of 1150 MHz.

A key technology implemented in these works is a microprocessor-controlled clutter-cancellation system, which creates an optimal signal to cancel the clutter from the rubble and the background (Chen et al., 2000a). The system is shown in Fig. 3.1. In this technique, a phase shifter and an attenuator were digitally controlled by a microcontroller to provide a delayed version of the transmitted signal. The delayed version of the transmitted signal was combined with the received signal using a directional coupler. Additionally, a microwave power detector was used to monitor the DC level of the combined signal, serving as the indicator for the degree of the clutter cancellation. The microcontroller automatically adjusts the phase delay and attenuation to minimize the DC level in the combined signal. And an optimal setting for the clutter cancellation corresponds to the point where the DC level of combined signal was minimized.

Having dealt with the problem of motion artifacts around the subject under test, the remaining problem is on the radar side. As the phase stability of the measurement system plays an important role in successful life signs detection, small unwanted mechanical motions of the transmit antenna result in unrecoverable phase errors in the received signals. To overcome this issue, it was proposed to use a bistatic radar with a sensor node receiver placed in the vicinity of the human subject (Mostafanezhad et al., 2007). The sensor node consists of an antenna and a mixer, similar to the radio frequency identification (RFID) tag used by Lubecke et al. (2002). It receives both the direct signal from the transmitter (LO) and the signal reflected from a human subject. Both signals are subject to the same "mechanical"

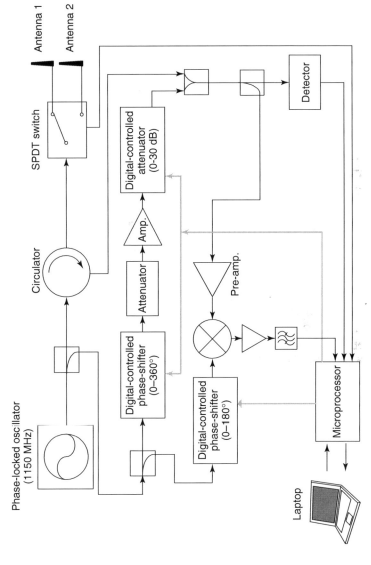

Figure 3.1 Block diagram of the clutter-cancellation system. Adapted from Chen et al. (2000a).

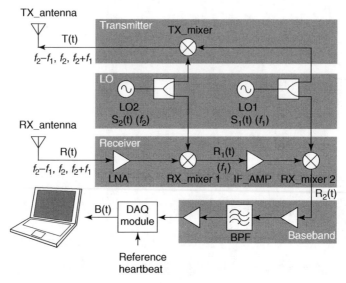

Figure 3.2   System block diagram of bench-top Ka-band radar (Xiao et al., 2006).

phase noise. If these path lengths are similar, there will be a significant phase noise reduction because of the range correlation effect, thus enabling accurate detection of life signs.

A common observation of microwave Doppler radar under small-angle linear approximation as discussed in Chapter 2 is that shorter wavelengths produce higher detection sensitivity. Therefore, a Ka-band double sideband bench top radar was built and reported by Xiao et al. (2006). Figure 3.2 shows the block diagram of the Ka-band radar. The receiver chain includes a receiving antenna (Rx_Antenna), a LNA, two down-converters (Rx_Mixer1 and Rx_Mixer2), and an IF amplifier (IF_AMP). The transmitter chain contains a transmitting antenna (Tx_Antenna) and an up-converter (Tx_Mixer). The baseband circuits contains a preamplifier (PreAMP), a band pass filter (BPF), and a low frequency amplifier (LF_AMP). The baseband output data is digitized by a data acquisition module (DAQ) that can also digitize a reference heart rate signal measured by another instrument, such as a piezoelectric type finger-tip pulse sensor or an electrocardiograph (ECG) for comparison. The digitized data can then be processed in a computer and the result can be displayed on the computer screen.

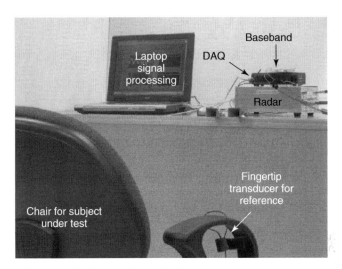

Figure 3.3 Setup of the Ka-band measurement system.

Two LOs generate signals $S_1(t)$ (with frequency f_1) and $S_2(t)$ (with frequency f_2). Two 3 dB power splitters are used to divide the power of $S_1(t)$ and $S_2(t)$, with half of the power sent to the transmitter chain and the other half sent to the receiver chain. A photo of the measurement setup is shown in Fig. 3.3.

As there is no filter following the Tx_Mixer, the output $T(t)$ of the Tx_Mixer has two main frequency components: lower sideband (LSB) $f_L = f_2 - f_1$ and upper sideband (USB) $f_U = f_2 + f_1$. An output power spectrum of the transmitter measured at antenna connector is shown in Fig. 3.4. The LSB and USB frequencies are 26.54 and 27.66 GHz with power levels of -6.21 and -9.63 dBm, respectively. The 27.10 GHz signal in between is the LO2 leakage due to nonideal port-to-port isolation of Tx_Mixer, and will be filtered out immediately after the first down-conversion. The other two second-order tones will be filtered out by the baseband filter too. It is shown in Chapter 2 that by using double sideband transmission, the radar can obtain reliable detection accuracy and avoid the null point problem by adjusting the IF oscillator.

The drawback of a Ka-band detector for vital sign detection, however, is noticed in experiments (Li et al., 2006a): when monitoring from the front of the body, the harmonic and intermodulation interferences to the heartbeat signal can prevent accurate detection at

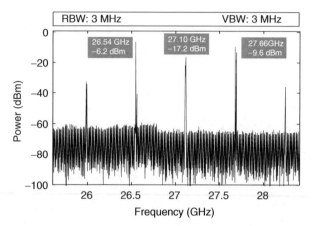

Figure 3.4 The output spectrum of the transmitter. The combined power in LSB and USB is equal to 350 μW power.

certain nonoptimum distances. Section 2.3.5 in Chapter 2 analyzes the mechanism of harmonic and intermodulation in details. Harmonics can interfere with desired heartbeat signal component and may affect detection accuracy, although, careful selection of optimal carrier frequencies for different subjects can minimize this effect (Li and Lin, 2007b). If used properly, though, harmonics can benefit certain new applications. This is the foundation for accurate measurement of both the amplitude and the frequency of periodic vibration using nonlinear Doppler phase modulation effect discussed in Chapter 2.

The above radar motion detection systems are all custom-built, which may be time-consuming and cost-inefficient for some research institutes to start research in related fields. Also, various considerations have to be taken into account for those radar systems. Laboratories have to spend significant time and efforts to design and build the radar system before they could actually conduct experiment. In reality, part of the research on radar motion sensor may focus on advanced signal processing and analysis techniques rather than the radar front-end itself. For example, the research of biometric identification by distinguishing an individual's radar vital sign signal carrying physiological and cardiopulmonary signatures is strongly dependent on signal processing. Therefore, a fast solution to assemble a motion detection system would be valuable for many research and prototyping activities. To realize a fast solution to build a radar motion system prototype, Gu et al. used classical RF instruments that are widely equipped in RF and communication labs (Gu et al., 2010). By connecting the instruments, the system is straightforward to realize. It can also be quickly disassembled for other applications. The system does not require any extra RF/microwave components and modules, neither does it need printed circuit board or integrated chip design. The system is easily configurable and very reliable, thus helps conduct system-level testing in early stage of research to acquire data and evaluate signal processing solutions.

The instrument-based radar system was built in reconfigurable radio architecture with digital I/Q demodulation technique. It includes the following instrument components: Agilent vector signal generator E8267C, Agilent spectrum analyzer E4407B, and Agilent vector signal analyzer 89600S. The simplified block diagram is illustrated in Fig. 3.5. The radar transceiver consists of E4407B, E8267C, and 89600S. The transmit antenna is connected to the vector signal generator E8267C, which acts as the source with a 20 GHz frequency tuning range. In the block diagram, the signal generator is simplified as a LO and a power amplifier (PA). A 10 MHz reference signal is generated as the clock for the whole system, so that all the components in the transceiver are synchronized, achieving coherent demodulation. The receiver chain consists of the RF front-end and the IF down-converter. The reflected microwave signal, modulated by the target motion, is received by the antenna and sent into the spectrum analyzer that down-converts the microwave signal to the 70 MHz IF output. This process is realized by the internal module Option H70 of E4407B, which automatically down-converts RF input to 70 MHz as long as the RF signal is lower than 26.5 GHz.

The down-converted received signal is then fed to the 89600S vector signal analyzer. The preselector shown in the dashed box of Fig. 3.5 has a tunable filter to reject out-of-band interferences that would otherwise create undesirable responses at the IF. The 89600S has digital signal processing features that provide full access to a wide range of vector signal analysis capabilities. Through GPIB connection, 89600S also controls the spectrum analyzer to fulfill coherent RF-to-IF signal down-conversion. The following standard components can be used in the vector network analyzer: 89601A vector signal analysis software with Option 100 (vector signal analysis), E8491B IEEE 1394 PC link

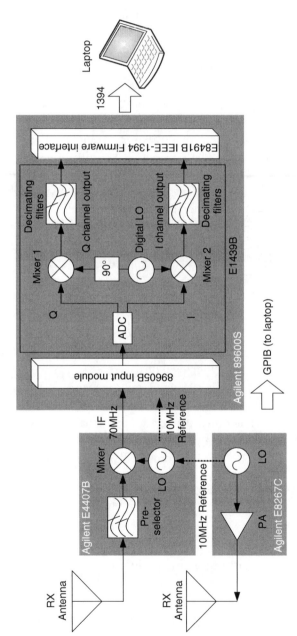

Figure 3.5 Block diagram of the instrument-based radar motion sensing system (Gu et al., 2010).

to VMEbus eXtensions for Instrumentation (VXI), C-size with Option 001 (OHCI-based IEEE 1394/PCI card), E1439 VXI 70 MHz IF digitizer, and 89605B RF input module with Option 611 (cable adapter kit). The down-converted 70 MHz IF signal is sent to the 89605B module that provides signal calibration. Digitization of the 70 MHz IF signal is carried out by the ADC E1439, with a sampling rate of 95 MSa/s and a bandwidth of 36 MHz. An anti-aliasing filter before digitization is used to limit the input frequencies of ADC to less than one-half of the sampling rate to avoid aliasing products.

The sampled data are stored and digital quadrature down-conversion is performed in the 89601A vector signal analysis software running externally on a computer connected to the analyzer via IEEE1394 PC Link. The digital demodulation in IF leads to very high demodulation accuracy because it assures accurate matching of the I and Q channels. Two decimating filters resample the I/Q outputs and generate digital baseband signals, which are recorded and processed in a laptop. The baseband digital output is then filtered, windowed, and auto-correlated in software such as MATLAB and LabVIEW. Spectral analysis is performed to extract the motion signal of interests.

Another approach to build up a CW radar motion sensing system is to use network analyzer. In (Obeid et al., 2012), a contactless microwave sensor was first realized using a vector network analyzer and two antennas for operations below Ka-band. With external oscillator and up-/down-converters, the system was further extended to 60 GHz. The system was able to detect heart beat variability at a distance of 1 m, using multiple operating frequencies of 2.4, 5.8, 10, 16, and 60 GHz.

3.1.2   Board Level Radar System Integration

Various board level radar motion sensor systems have been used in different applications. Most of the board level systems use off-the-shelf surface-mount technology (SMT) components to realize the radar front-end as well as other necessary functions. The convenience for low power portable application is one of the driving forces for the board level development. As an example, Fig. 3.6 presents a software-defined multifunction radar motion sensor that enables multiple detection modes for optimized performance in different applications.

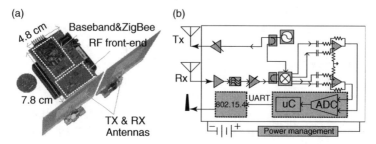

Figure 3.6   (a) A topside view and (b) the block diagram of the 2.4 GHz multifunction interferometric radar sensor. The radar sensor is integrated with a microcontroller and a ZigBee node (Gu et al., 2012b).

The sensor was fabricated on Rogers 4350 laminate and has a dimension of 7.8 cm × 4.8 cm. Two 2.4 GHz patch antennas were designed for the radar sensor, with one for transmitting and the other for receiving. A ZigBee module was integrated on top of the sensor board. The radar sensor is either powered by an external power source or by a 4.5–6.0 V battery pack lying under the sensor board. The input power is regulated by the power management unit on the sensor to feed different parts of the sensor circuit. A free running VCO generates a 2.4 GHz unmodulated single-tone signal, which is divided equally into two parts by a balun. One part of the signal is amplified and transmitted out via the patch antenna. The other part goes to the demodulator as the LO signal. The receiver chain receives the backscattered signal with motion information modulated in the phase. The received signal is amplified, filtered, and goes into the quadrature direct-conversion demodulator that mixes the received signal with the LO signal. The baseband output of the demodulator is further amplified and digitized by a 10-bit ADC module in the micro-controller.

The micro-controller processes the digitized signals and then sends them to the ZigBee module that connects to a corresponding ZigBee mesh network for wireless data communication. The ZigBee module allows the radar sensor to act as an end device in the ZigBee mesh network. The radar sensor could establish communication to a mesh network very quickly by verification of a network ID number. This makes it easy to form a sensor network for monitoring different parts of large structures. The integrated micro-controller allows multiple functions running on the same platform. Different signal processing schemes can be realized. For example, the radar sensor could work in the complex signal demodulation, arctangent demodulation, nonlinear demodulation, or small angle approximation mode. The multi-function is useful because it could adjust the radar sensor to operate with optimal performance in different environments.

3.1.3 Motion Sensing Radar-On-Chip Integration

3.1.3.1 Early Radar-On-Chip Development. The early radar-on-chip integration of motion sensing function started from noncontact vital sign monitoring applications. The first integrated chip for noncontact vital sign detection was developed at Bell Labs during 2000–2002. In the late 1990s, researchers in Bell Labs began the research work to integrate Doppler radar sensing function in cell phones and other portable wireless communications devices as a means to detect the user's heartbeat and respiration (Lubecke et al., 2002; Boric-Lubecke et al., 2003). Using the BiCMOS chip set developed for cellular base station RF transceiver front-ends (Droitcour et al., 2004b), an integrated physiological motion sensor radar printed circuit board (PCB) module was developed and reported in 2001 (Droitcour et al., 2001). The chip set includes an LNA, a double-balanced resistive mixer, a VCO, and an active balun buffer amplifier. Although the chip set was optimized for low noise and high linearity for the purpose of shrinking cellular network base station's RF front-end, it was found suitable for building physiological motion sensor radar. Fully integrated vital sign detection sensor chips, implemented in 0.25 μm BiCMOS and CMOS processes, were later developed and reported in 2002 (Droitcour et al., 2002); this marked the first demonstration of the noncontact physiological motion sensor chip. It showed the feasibility of making these sensors in large quantity and low cost, and the potential of integrating them into portable electronic devices. The photograph of a sensor chip is shown in Fig. 3.7 (Droitcour et al., 2003). The chip, operating at 2.4 GHz, was implemented in 0.25 μm CMOS process and had direct-conversion quadrature receiver architecture. The chip was packaged in a TQFP 48-pin package. The chip delivered 2 mW at RF output and dissipated 180 mW from DC supply.

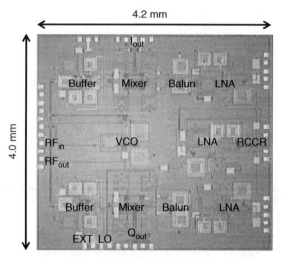

Figure 3.7 The first noncontact physiological motion sensor chip using quadrature receiver architecture (Droitcour et al., 2003).

3.1.3.2 24 GHz Intruder Detection Radar Sensor Chip.

Taking advantage of the same Doppler Effect, integrated 24 GHz intruder detection radar (Morinaga et al., 2007) has been developed by the Hitachi central research lab to cover a fan-shaped ground area with $90°$ in azimuth and 0 to over 14 m in range, when mounted at a height of 5 m. The radar was implemented with three vertically switched-beam antennas; each switched beam monitors a different range segment with a mono-pulse scheme employed to achieve wide azimuth coverage. Figure 3.8 shows the 24 GHz radar block diagram and its separate detection areas monitored with three antennas.

The performance of a prototype radar was experimentally evaluated on a PC-based platform used for RF module control, data collection, and signal processing. Two discrete frequencies with a separation of 2 MHz were transmitted inside a 150 μs interval during experiment. This leads to an unambiguous maximum range measurement of 37.5 m. As the radar can detect multiple portions of a single human body moving at different velocities, an S/N (signal-to-noise)-weighted arithmetic mean of two-dimensional positions of the body parts is calculated to locate a human target. The boundaries of the monitoring areas covered by the three pairs of antennas were set to 4 and 8 m with

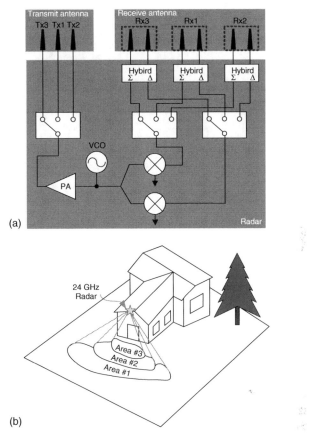

Figure 3.8 (a) Block diagram of the 24 GHz intruder detection radar, and (b) the separate detection areas monitored with three antennas. Adapted from Morinaga et al., (2007).

1 m overlapping in the Y-direction (as shown in Fig. 3.8) where the gains of two adjacent antennas are the same. Experiment results show that the radar integrated in monolithic microwave integrated circuit (MMIC)'s can successfully detect a human intruder with a position accuracy of 50 cm when moving at 1.4 m/s.

3.1.3.3   5 GHz Indirect-Conversion Double-Sideband Radar Sensor Chip.   As the indirect-conversion radio architecture can avoid the null detection point by tuning the IF, a 5 GHz double-sideband radar sensor was subsequently designed and fabricated using a UMC (United

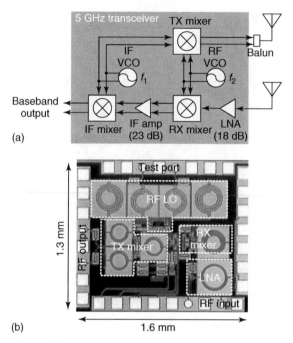

Figure 3.9   Block diagram (a) and chip micrograph (b) of the 5 GHz radar transceiver designed and fabricated using an United Microelectronics Corp (UMC) 0.18 μm mixed signal/RF process (Li et al., 2008b).

Microelectronics Corporation) 0.18 μm mixed-mode/RF CMOS process (Li et al., 2008b). The block diagram and chip micrograph of the resulting radar sensor chip are shown in Fig. 3.9. The transmitter simply uses two VCOs as signal sources and a Gilbert double-balanced mixer as the up-converter (Up-C). The receiver chain has a LNA, a Gilbert double-balanced mixer as the down-converter (Down-C), an IF amplifier (IF-Amp), and a second stage passive mixer (IF-Mixer).

Measurements showed that this chip has a tuning range over 1 GHz; it was also shown that the differential architecture has the advantage of reducing LO leakage. The on-chip integrated 5 GHz radar was tested in a lab environment for vital sign detection of a human subject 1.75 m in height. Measurements were performed when the subject was seated at 0.5, 1, 1.5, and 2 m away from the radar, facing the antenna. As shown in Fig. 3.10, both the respiration and heartbeat components were successfully detected in the baseband spectrum.

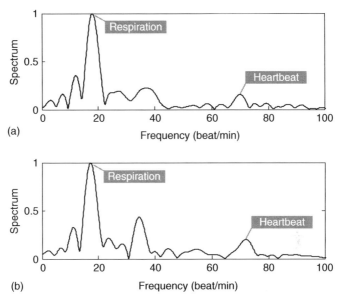

Figure 3.10   Baseband spectrum detected by the on-chip integrated 5 GHz radar when the subject was 0.5 m away from the antenna (a) and 2 m away from the antenna (b).

3.1.3.4   *High Sensitivity Direct-Conversion Radar Sensor Receiver Chip.*   In Chapter 2, the special gain and noise budget consideration for high-sensitivity motion sensor was derived. For high sensitivity detection of mechanical motion, large gains in the RF blocks (LNA + preamplifier) and low baseband flicker noise (mixer + VGA) are emphasized. On the basis of this, a direct conversion 5–6 GHz radar sensor receiver chip was designed. The block diagram of the chip is shown in Fig. 3.11. To help readers gain insights of implementing a micro-radar sensor on a semiconductor chip, the rest of this section describes the details of the design and measurement results.

The receiver chain consists of a LNA, a gain adjustable preamplifier (PreAmp) to further amplify the signal before down-converting to baseband, two passive mixers for I/Q channels, and two VGAs. A Gm-boosted bias circuit combining constant-Gm and bandgap references performs temperature compensation for the LNA and preamplifier. A bandgap voltage reference circuit is used to bias the mixer at the minimum flicker noise operation state. The VGAs are biased by

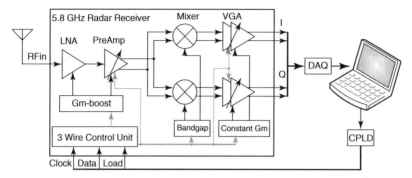

Figure 3.11   Block diagram of the high sensitivity direct-conversion radar sensor chip (Li et al., 2010b).

a constant-Gm circuit. A 3-wire digital program/control unit is integrated on-chip to decode the digital control codes. The control bits can be programmed into the chip to set the operation point and receiver gain. A complex programmable logic device (CPLD) board is used to communicate with the chip.

Applying the two-way Radar Range equation to noncontact motion sensing, the received signal power P_r varies as the distance R between the subject and antenna changes

$$P_r = P_t G_t G_r \frac{\lambda^2 \sigma}{(4\pi)^3 R^4}, \tag{3.1}$$

where P_t is the transmitted signal power, G_t and G_r are the gains of the transmit/receive antenna, σ is the radar cross section, and λ is the wavelength. The received signal power is reduced by 12 dB as the distance between the radar detector and the subject doubles. A two-stage digitally controlled VGA is implemented to provide desired dynamic range.

On the other hand, the unlicensed 5 GHz band is popular for many wireless applications. The noncontact vital sign detector is designed for civilian applications with potential radio interferers. It is necessary to avoid in-band interference from saturating the RF front-end, especially the mixer. Therefore, the preamplifier between the LNA and the mixer was designed to have high/low gain modes with different input-referred 1 dB compression point ($P1dB$).

Another issue of noncontact motion sensor is the interference by the harmonics of the signal components. For example, when a motion sensor is configured for vital sign monitoring, the third-order or forthorder harmonic of respiratory signal will have frequencies that are close to the heartbeat signal, which may corrupt the detection of heartbeat signal. The harmonics are generated by the nonlinear Doppler phase modulation and the nonlinearity of the electronic circuit. Differential circuit can effectively reduce the even-order harmonics. Therefore, the preamplifier was also designed as an active single-to-differential balun.

On the basis of the above considerations, the performances of the building blocks and receiver chain were optimized for noncontact vital sign detection application, as listed in Table 3.1. Figure 3.12 shows the simplified schematics of the receiver chain without showing bias circuits and control units. The detailed circuit design is discussed as follows.

As discussed in Chapter 2, it is desirable to have a wide RF tuning range because the optimal carrier frequency varies for motions with different amplitudes. Resistors in series with the loading inductors are used in the LNA and preamplifier to broaden the RF output bandwidth. At the input of the LNA, a pad with large parasitic capacitance (210 fF) forms a T-matching network with the RF bonding wire L_b and the on-chip inductor L_g, achieving an input bandwidth of larger than 1 GHz. Because single-ended antennas are used in the radar system, a single-ended cascode LNA instead of differential LNA is used to avoid off-chip single-to-differential conversion module.

The preamplifier is designed for two purposes. First, it acts as an active single-to-differential balun. Second, it provides low/high gain options to extend dynamic range. When large in-band interference is present or the subject is close (<0.5 m) to the antenna, the preamplifier switches to the low-gain mode with high input $P1\text{dB}$ of 0.5 dBm. To realize this, two cascode amplifier stages are placed in parallel to drive the output load. The switch between the high/low gain cascode is controlled by bits D_1 and D_2 in Fig. 3.12.

Feedback capacitors C_{b1} and C_{b2} are used in the cascode amplifier to produce differential output. The mechanism is illustrated in Fig. 3.13a. Single-ended input voltage v_{in} is applied to the gate of transistor M_a, which induces AC current i on the left branch of the differential pair. This current is divided among parasitic capacitance C_{ST},

TABLE 3.1 Performances of the Building Blocks and Receiver Chain (Li et al., 2010b)

RX Performance	Current, mA	Gain, dB	IIP3, dBm	P1dB, dBm	NF, dB	BW, GHz	Note
LNA	4.6	24.6	—	>−28	2.56	5.3–6.3	Post layout
PreAmp (low gain)	10.3(3.3 + 7)	0.2	—	0.5	17.12	Broad	Post layout
PreAmp (high gain)	11(4 + 7)	11.5	—	−9.5	7.85	Broad	Post layout
LNA + preAmp low	14.9	24.4	—	−29.8	2.73	5.3–6.2	Post layout
LNA + preAmp high	15.60	35.9	—	−39.2	2.59	5.3–6.2	Post layout
Mixer + buffer	0.5 × 2	−5.0	12.45	2.84	69.14 @ 1 Hz	Broad	Post layout
VGA	(0.4 + 0.8) × 2	18/24/30/36	See VGA compliance matrix			–	
RX MinGain	24 (26)	37 (36)	—	−32 (−34.5)	42.03 dB @ 1 Hz 56.9 nV/sqrt(Hz) @ 1 Hz	~1 GHz (0.8 G)	Simplified (Measured)
RX MaxGain	26 (28)	67 (63)	—	−64 (−60)	29.98 dB @ 1 Hz 14.36 nV/sqrt(Hz) @ 1 Hz	~1 GHz (0.8 G)	Simplified (Measured)

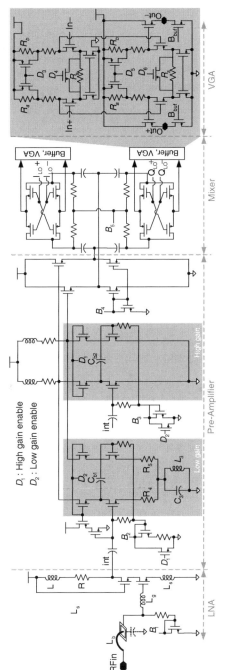

Figure 3.12 Simplified circuit schematics of the receiver chain (Li et al., 2010b).

71

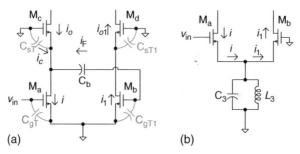

Figure 3.13   Two mechanisms of single-to-differential conversion in preamplifier: (a) capacitive feedback; (b) common-source impedance boosting with LC tank.

the impedance looking into the source of M_c, and the total capacitance of the feedback path (i.e., series connection of feedback capacitor C_b and gate-associated parasitic capacitance C_{gT1}). The feedback current i_F produces an AC voltage on the gate of M_b, resulting in an AC current i_{o1} on the right branch of the differential pair. By properly adjusting transistor size and C_b, i_o, and i_{o1} can be equalized, realizing single-to-differential conversion.

For the high gain cascode, a fine-tuned C_{b2} can achieve better than 0.4 dB amplitude imbalance over process–voltage–temperature (PVT) variations. For the low gain cascode, however, it is difficult to achieve desired performance with only a feedback capacitor because of the low current gain of the input transistor. Therefore, a parallel LC tank ($L_3 C_3$) is implemented to provide high resonant impedance at the source of the low gain cascode. This mechanism is illustrated in Fig. 3.13b, where the LC tank provides high impedance at the common source node, directing the source current of M_a into M_b. In real implementation, source degeneration resistors R_4 and R_5 are also used to improve the linearity of the low gain cascode stage, as shown in Fig. 3.12.

A source follower buffer is used to drive the passive mixer in voltage-driven mode as the mixer uses large transistors to achieve better noise performance, as discussed in Section 3.1.3.5. The source follower was chosen because it improves the isolation and is robust against load variation and process-voltage-temperature variations.

The bias of the RF front-end is very important for the performance of the radar chip. The LNA and preamplifier share the same

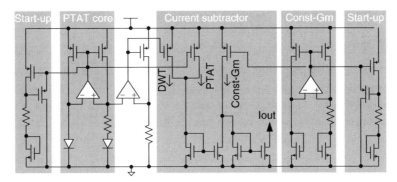

Figure 3.14 Gm-boosted bias circuit for the LNA and preamplifier.

Gm-boosted bias circuit. A straightforward choice is the classic constant-Gm circuit, whose transconductance depends on the reference resistor only. However, the sheet resistances of on-chip resistors also have a temperature coefficient and the available g_m is limited at high temperature. To boost the transconductance at high temperature, bandgap, and constant-Gm circuits are used simultaneously in the bias circuit of Fig. 3.14, and a current subtractor is used to subtract a decreased-with-temperature (DWT) current from the current of the constant-Gm circuit. In the meantime, a small amount of proportional to absolute temperature (PTAT) current was used to optimize the output curvature. As a result, the transconductance corresponding to the output reference current is boosted to achieve better temperature performance for the radio front-end.

The flicker noise is very high around the baseband signal frequency, posing a challenge for the noise performance of the mixer in the direct conversion receiver. Conventional Gilbert-Type active mixers contain several noise sources: the transconductor noise, the LO noise, and the noise from the switching transistors. These noise sources establish an unacceptable noise figure in the interested baseband spectrum. Therefore, Gilbert-type active mixers are not suitable for the vital sign detection receiver. Passive mixers avoid the transconductor stage in the active mixers and have no DC bias current. This feature minimizes the flicker noise at the mixer output. Therefore, a passive mixer was chosen for direct conversion radar receiver.

To minimize the flicker noise of the passive mixer, the gate-source voltage of the switching transistor is close to V_{th} in a break-before-make mode, which minimizes the DC bias current in the switching transistors. For the switching transistor size, there is a trade-off between noise figure and capacitive load to the preamplifier stage. To provide appropriate noise figure in the interested vital sign bandwidth, large switching transistors are used and they produce relatively large capacitive loads to the preamplifier. As a result, there is a large capacitive load to the preamplifier stage. This is the reason that a source follower buffer was used at the preamplifier to drive the mixer in voltage-driven mode. The LO signal power is set to 1 dBm for maximum gain based on simulation.

According to Nguyen et al. (2006), a voltage-driven mode passive mixer does not have as good linearity as a current-driven mode passive mixer. In order to relax the specification of the mixer linearity, the preamplifier was designed to be gain and linearity programmable. A bandgap bias circuit was designed for the mixer.

The flicker noise and the baseband bandwidth are two important issues for the baseband VGA design. To reduce the flicker noise, large differential pairs ($W = 180$ μm, $L = 3$ μm) with resistor loads are used. The transistors also provide large parasitic capacitance to increase the time constants of mixer and VGA, so that the baseband output 3-dB bandwidth of the whole receiver chain is limited to 1 MHz without the need of any off-chip components. It should be noted that the bandwidth could be further reduced to limit the receiver white noise for some applications. 1 MHz bandwidth was chosen so that the receiver chip could be used for other purposes such as measuring fast mechanical vibration.

In order to increase the total gain of the receiver chip, a two-stage VGA shown in Fig. 3.12 was designed to provide five gain steps ranging from 12 to 36 dB. The maximum gain is obtained when D_3 to D_6 are all high. When D_4 or D_6 is low, R_s degenerates the differential-mode signal and reduces the gain by approximately 6 dB, while improving the input P1dB by 7 dB. When D_3 or D_5 is low, the load resistors R_a and R_b are shorted and the gain is further reduced by 6 dB for each stage. In experiment, the VGA outputs are sampled by a 12-bit DAQ (NI-6008) through the differential sampling channels. A source follower is used as a buffer and level shifter at the VGA output. The VGA input-referred noise voltage is designed to be less than 708.2 nV/sqrt(Hz) at a frequency of 1 Hz. The VGA specifications at different programmed gain stages are shown in Table 3.2.

TABLE 3.2   VGA Control Code and Specifications (Li et al., 2010b)

	VGA Specification Compliance Matrix				
Code $<3:0>$	Simplified Gain	Simplified $P1$, dB	NF, dB @ 1 Hz	V_{in}, nV/sqrt(Hz)	Note
0000	12.66	−6.1	64.0	708.2n	Backup
0100	18.89	−13.2	62.9	624.8n	
0110	24.94	−20.75	62.7	607.9n	Not used
0111	30.71	−26.31	62.6	606.1n	Not used
1100	24.8	−19.19	57.33	328.9n	
1110	30.85	−26.65	57.11	320.7n	
1111	36.62	−32.18	57.08	319.8n	

Code = st1$<1:0>$, st2$<1:0>$.
DAQ rms quantization error: 0.352 mV.

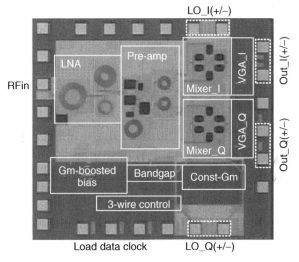

Figure 3.15   Chip microphotograph of the radar sensor receiver (1.2 mm × 1.2 mm) (Li et al., 2010b).

The radar sensor receiver chip was fabricated in UMC 0.13 μm process. Figure 3.15 shows the microphotograph of the fabricated chip. A circuit board with a Xilinx XC9536 CPLD was built to program the chip at 10 Mbps, minimizing the configuration time.

For pure electronic characterization, two synchronized signal generators were used to generate the receiver input signal and mixer's LO

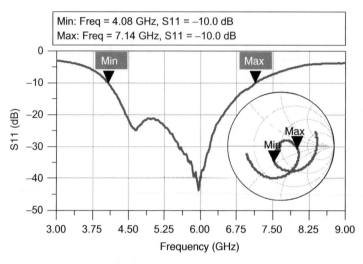

Figure 3.16   Measured input $S11$ of the receiver chip.

drive. The signal generators have a 1 Hz frequency offset to imitate the effect of detecting a heartbeat signal. At the baseband output, an oscilloscope was used to record the output at a sampling rate of 1 kHz with a record length of 10,000 data points. The recorded digital signal can be resampled with a lower ADC sampling rate and resolution to obtain the performance of a portable DAQ used in practical vital sign detection. In this way, the effect of different ADC sampling rate and resolution can be compared. Other specifications such as voltage gain, $P1dB$, and bandwidth can be obtained based on the recorded baseband signal. The overall performance of the fabricated chip is compared with simulation results and shown in Table 3.1.

The input $S11$ of the receiver chip is measured using an Agilent E8361A vector network analyzer as shown in Fig. 3.16. An input matching bandwidth from 4.1 to 7.1 GHz was realized.

The single-to-differential conversion function of the preamplifier has been verified at the baseband output. Figure 3.17 shows the measured I-channel baseband output with a -83 dBm 5.8 GHz input RF signal and an LO signal of 1 Hz offset driving the down-converter. As shown in Fig. 3.17a, the positive and negative outputs have negligible amplitude and phase mismatch, verifying the effectiveness of the preamplifier with feedback capacitors for single-to-differential conversion. Figure 3.18 shows the corresponding spectrum of the

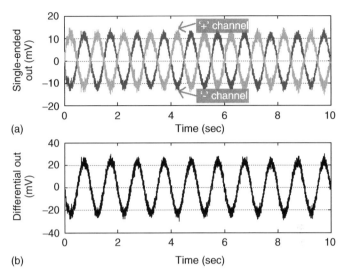

Figure 3.17 Baseband time domain output when input signal is −83 dBm. (a) Positive and negative outputs of the *I* channel, (b) *I*-channel differential output.

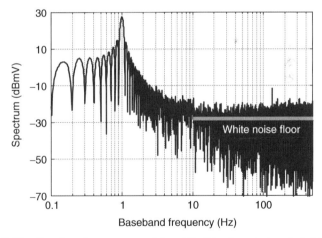

Figure 3.18 Baseband spectrum with −83 dBm input signal, 1 kHz sampling rate, and 0.2 mV quantization resolution. The average white noise floor is indicated in the figure and measured as −28.02 dBmV.

I-channel baseband output. The white noise floor was obtained by averaging the noise over frequencies higher than 10 Hz, and was found to be −28.02 dBmV as indicated in the figure. It should be noted that although the receiver flicker noise corner frequency is far above 10 Hz, the baseband output noise is quickly dominated by white noise because of the aliasing and folding of baseband white noise into the sampling frequency. The measurement results verify that for a CMOS receiver chip, the flicker noise can be suppressed by proper design based on noise considerations.

The high sensitivity was tested by feeding a weak signal into the receiver chip. Figure 3.19 shows the *Q*-channel baseband output when the RF input was reduced to −101 dBm and a different baseband ADC sampling rate was used. Figure 3.19a shows the baseband differential output with a sampling rate of 1 kHz, while Fig. 3.19b shows the same baseband signal obtained with a sampling rate of 20 Hz. The 20 Hz

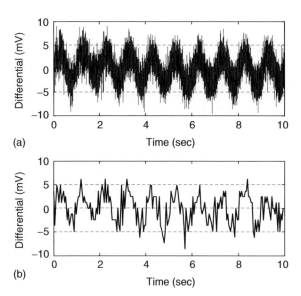

Figure 3.19 *Q*-channel differential output when input signal is −101 dBm. (a) Sampling rate is 1 kHz, quantization resolution is 0.2 mV. (b) Sampling rate is 20 Hz, quantization resolution is 1.2 mV, corresponding to the real case of vital sign detection using NI-6008 USB data acquisition module for A/D conversion.

sampling rate was chosen because it is the sampling rate often used in lab environment for low cost portable vital sign detection applications.

Figure 3.19 and Fig. 3.20 show the baseband spectra of detected signals with sampling rates of 1 kHz and 20 Hz, respectively, for the same −101 dBm input signal. The average white noise level is also measured and indicated in the figures. By comparing the result of Fig. 3.20 and Fig. 3.18, it shows that although the input signal was reduced by 18 dB, the white noise stays at the same level of around −28 dBmV. Furthermore, when comparing Fig. 3.20 and Fig. 3.21, it is shown that the signal-to-noise level improves by 17 dB as the sampling frequency was increased from 20 Hz to 1 kHz. This demonstrates the importance of ADC sampling rate on receiver sensitivity: the system performance could be largely improved by over-sampling. It should be noted that, even in the worst case using a low sampling rate of 20 Hz, the SNR on the periodogram is around 20 dB for the −101 dBm input signal. Therefore, the receiver sensitivity is better than −101 dBm in the absence of clutter noise and other undesired motion-induced noise.

The receiver input $P1dB$ was measured and shown in Fig. 3.22, when the receiver was programmed in the highest gain mode and the lowest gain mode. As the receiver is configured from the high gain to the low gain mode, the measured input 1 dB compression point increases from −60 to −34.5 dBm, while the gain drops from 63 to 36 dB. When the receiver was configured into high gain mode, the accumulated DC offset along the receiver chain affects the balance and DC operation point of the differential pair, making the measured gain lower than the simulated gain. A digital DC offset calibration loop can be added into the receiver chip to eliminate the DC offset.

To demonstrate motion detection function, experiments have been performed in a lab environment to detect the vital signs of seated people. Figure 3.23 shows the experimental setup. The sensor chip was mounted on a Rogers printed circuit board and powered by a commercial 1.5 V battery. Two by two patch antenna arrays were used to transmit and receive signal. For direct conversion vital sign detection, the design challenge is on the receiver side. The transmitter only requires a simple single-frequency CW signal source. A microwave circuit board was designed to split the signal from a signal generator, producing the transmitted signal and two quadrature LO drives for mixer. The microwave circuit board has a Wilkinson power divider, a

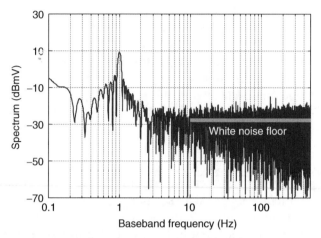

Figure 3.20 Baseband spectrum with −101 dBm input signal, 1 kHz sampling rate, and 0.2 mV quantization resolution. The average white noise floor is indicated in the figure and measured as −27.93 dBmV.

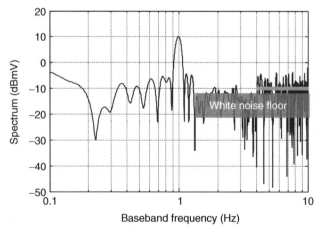

Figure 3.21 Baseband spectrum with −101 dBm input signal, 20 Hz sampling rate, and 1.2 mV quantization resolution. The average white noise floor is indicated in the figure and measured as −9.89 dBmV.

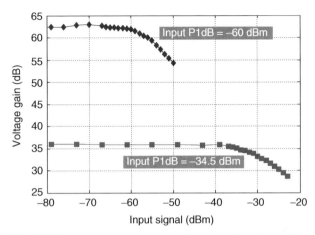

Figure 3.22 P1dB measurement result when receiver is configured to the highest and the lowest gain modes.

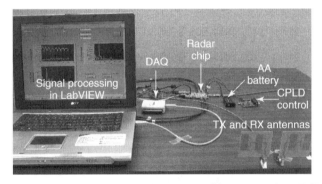

Figure 3.23 Experimental setup with sensor chip mounted on PCB, CPLD board, TX/RX antennas, DAQ, battery, and the LabVIEW interface (Li et al., 2010b).

branch line coupler, and two rat-race couplers. The signal fed into the transmitting antenna was 7 dBm, and mixer LO drive was 1 dBm.

Without any external baseband amplifier, the radar sensor chip can detect the vital sign of human subjects at a distance of up to 3 m. The detection range can be further increased if higher transmitted power or a DAQ with higher sampling rate is used to increase SNR.

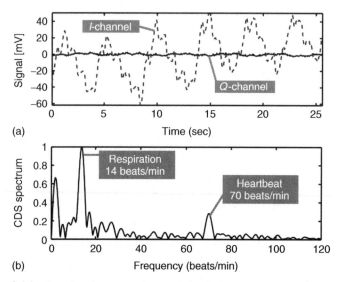

Figure 3.24 Baseband output detected from the back of a human subject. (a) Time-domain waveform of the I (dashed line) and Q (solid line) channels. (b) Complex signal demodulated (CSD) spectrum.

Because the radar chip has I/Q channels, complex signal demodulation discussed in Chapter 2 was used for baseband signal processing. Figure 3.24 shows an example of baseband time-domain signals and spectrum detected from the back of a human subject. Clean time domain signals were observed. The detection was performed at a distance where the I channel is near the optimal detection point while the Q channel is close to the null detection point. Because of the complex signal demodulation, the spectrum was not affected by the presence of a null detection channel. The respiration and heartbeat signal components can be identified from the spectrum, indicating a respiration rate of around 14 times/min and a heartbeat rate of around 70 beats/min.

It is interesting to observe that the time domain waveforms present less noise than those measured by the 5 GHz double sideband transceiver chip. This can be explained by the improved design based on the sensitivity analyses and the fully integrated on-chip bias and amplifier circuitry presented in this paper. Table 3.3 summarizes the radar detection range and the related optimal gain settings for the preamplifier and VGA.

TABLE 3.3   Radar Detection Range and Gain Settings (Li et al., 2010b)

Operation Range Versus Gain Step Settings					
Distance, m	<0.5	0.5	1	1.5	2	>2
Received power, dBm	>−37	−37	−49	−56	−61	<−61
PreAmp gain, dB	Low	Low	High	High	High	High
VGA gain, dB	18	18	18	24	30	36

3.1.3.5   60 GHz High Sensitivity Radar-On-Chip Solution.   Operating the vital sign detection radar at millimeter-wave frequency and above presents a challenge for detecting both respiration and heartbeat. This is because the respiration displacement is comparable to the wavelength and the heartbeat displacement is usually much smaller. In this case, the harmonics of respiration signal due to nonlinear modulation complicates the spectrum, and the sensitivity may no longer increase with decrease in wavelength (Li and Lin, 2007b). Nevertheless, the small size of the antenna and the potential of vital sign imaging using phased array technique still draw interest. By using UMC 90 nm CMOS process and flip-chip packaging to integrate with the antennas, a system-in-package micro-radar for vital sign detection was demonstrated (Kao et al., 2012). The micro-radar operates at 60 GHz unlicensed band and uses heterodyne architecture with quadrature demodulation at IF = 6 GHz. Although, the propagation of 60 GHz signal in air experiences high loss due to oxygen absorption, this unlicensed band is suitable for short range vital sign detection.

The system architecture of the radar transceiver chip is shown in Fig. 3.25. A picture of the micro-radar indicating the locations of the radar transceiver chip and the antennas is also shown. The complete micro-radar hardware is small and weighs less than 10 g. The total power consumption of RF, IF, and baseband circuits is 377 mW from 1.2 V supply. The 60 GHz micro-radar can detect vibrations as small as 20 μm from 0.3 m away and can certainly detect respiration displacement, as well as heartbeat while the subject is holding the breath. The challenge arises when trying to detect both respiration and heartbeat because the large respiration displacement introduces strong nonlinear phase modulation effect that interferes with the detection of small heartbeat displacement. Further, signal processing technique is in need to solve this issue.

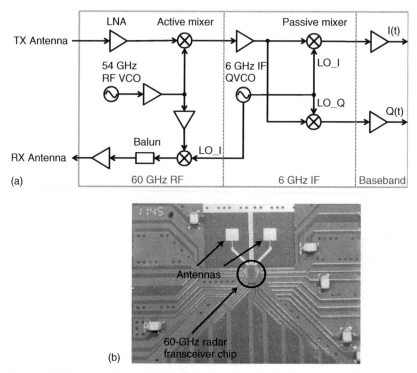

Figure 3.25   60 GHz microradar. (a) System block diagram. (b) Photograph of the integrated radar.

Other millimeter wave frequency Doppler radars have also been reported for the measurement of periodic movement including vital signs (Huey-Ru et al., 2012; Petkie et al., 2009; Bakhtiari et al., 2012; Mikhelson et al., 2012b). In Petkie et al. (2009), a 228 GHz carrier was used for noncontact vital sign detection because this frequency is in an atmospheric window with at least 50% single-pass transmission. In addition, higher frequency can maintain a collimated beam over much greater distances for reasonable aperture sizes, and the radar cross section of the vital sign area may also increase as frequency increases. This 228 GHz system has successfully extended the respiration and heart rate measurement to a range of 50 m. The encouraging results verified that long-range detection is feasible by using higher carrier frequency.

3.1.4   Pulse-Doppler Radar and Ultra-Wideband Technologies

3.1.4.1 UWB Motion Sensing for Vital Sign Detection. Pulse-Doppler radar and ultra-wideband (UWB) radars can be used in many cases of highly precise remote observation of moving objects at short distances. During detection, the EM wave pulses radiated by radar are scattered by a moving object. Owing to the Doppler Effect, the oscillation frequency within the pulse will be changed. As frequency variation leads to variations of the oscillation period while the total number of periods in the pulse remains the same, the duration of a scattered pulse also changes. Owing to the same effect, the repetition frequency of pulses scattered by an object will also change. The sign of these variations depends on the direction of target movement relative to the radar. The variation value depends on the object's motion speed. Therefore, the signal spectrum will be changed either within the bandwidth or at the center frequency.

The overall concept of UWB motion sensing radar has been discussed by Staderini (2002). A straightforward application in the detection of heart and breathing activities was illustrated. It was also claimed that a UWB radar could replace presently used fetal monitors based on ultrasound and pressure sensors by providing data about maternal heart rate, maternal breath rate, fetal heart rate, fetal movements, and uterine contractions. The advantage of providing convenient uninterrupted mother and child care has been discussed. As a simple example, a UWB radar prototype realized in 1998 at "La Sapienza" University of Rome was demonstrated. It was a breadboard version of UWB radar operating around 800 MHz. It uses commercially available components and detects heart and body movements of a human target. Dipole antenna was used to transmit and receive the broadband radar signal.

An important idea brought by Staderini is the optical UWB radar, in which the receiving and transmitting antennas are replaced by a laser-diode and photo-diode pair. Instead of emitting a short EM pulse, a short train-wave of light will be emitted and the echoes will be detected by a very fast PIN photodiode. As the human skull and brain can be approximated as transparent to infrared (IR) energy and the IR transmission spectra of the whole brain are used to investigate brain metabolism and blood circulation, it inspires significant potential biomedical applications. The same studies as UWB radar-based systems, such as the time-resolved technique, can be performed to enable IR imaging of sectors of the brain *in vivo* in a noninvasive fashion. One particular area of great interest for the "optical UWB radar" is the estimation of hemoglobin oxygen saturation in selected brain districts.

On the further development of UWB motion detection radar, Immoreev, Samkov, and Tao reported short-distance UWB radars in Russia and Taiwan. In 2005 (Immoreev et al., 2005), the UWB radar has achieved an operation range of 0.1–3 m, with a transmitted pulse power of 0.4 W. The duration of pulses exciting the transmitting antenna was 300 ps, and the repetition frequency is 2 MHz. The central frequency in a radiated pulse's spectrum is around 1 GHz. The radar achieves a receiver sensitivity of −77 dBm and a dynamic range of 34 dB. The range of Doppler frequencies of extracted signals is 0.16–40 Hz. The radar was successfully applied in medical application to monitor heart and respiratory beats. The variability of cardiac rhythm obtained by this radar was compared with ECG's data. Experimental results confirmed the potential of using UWB radar for monitoring human's heart activity. The same radar was also demonstrated for detection of live people concealed behind nontransparent barriers, as well as people crossing a guarded perimeter line.

In 2008, Immoreev and Tao (2008) demonstrated a new version of UWB radar for patient monitoring. The radar achieved an operation range of 0.6 to 3.5 m using a signal spectrum from 6.2 to 6.6 GHz and a signal spectrum from 5.75 to 7.35 GHz. The pulse repetition frequency was 2 MHz. The radar has a radiated beam width of 31.5°. It consumes a DC power of 1 W. The instantaneous pulse power was 9 mW, and the average power was 0.05 mW. It meets the Federal Communications Commission (FCC) requirements. In every pulse repetition period, the pulse generator forms two short UWB pulses. The first short pulse from the generator enters a broadband amplifier and is fed to the antenna to be transmitted to the target. After that, the second short pulse from the generator is sent to the reference channel of a quadrature phase detector. The functional block diagram of the system is shown in Fig. 3.26. The radar antenna captures the pulse reflected from the target. Then the received pulse is amplified by an LNA and sent to the quadrature phase detector, which detects the phase change that reflects the physiological motion. The phase detector output is further amplified, digitized by an ADC, and sent to a computer for

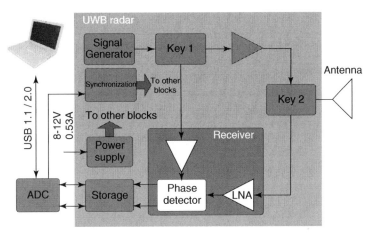

Figure 3.26 Blocked diagram of the UWB radar. Adapted from Immoreev and Tao (2008).

signal processing. The device has been successfully tested in Moscow and Taiwan for medical applications.

3.1.4.2 *UWB Range Detection and See-Through-Wall Radar.* An important advantage of UWB radar is the range detection capability. On the basis of this, a real-time see-through-wall radar system was developed and implemented by Yang and Fathy (2009). In that work, a Vivaldi array was adopted for the transmitting and receiving antenna. A custom low-cost module based on commercial field-programmable gate array (FPGA) boards and low-speed ADCs were used for data acquisition. The introduced module does not require a custom implementation of high-speed wideband mixed-signal circuitry but only depends on the FPGA firmware design, which favors a rapid system prototyping. The DAQ accomplished a 100 ps equivalent-time sampling resolution at 100 Msps, while the developed system provides a 2-D real-time view of motion with a 1.5 ms speed behind walls. The system allows for an easy reconfiguration to support multiple operating frequency ranges, array deployments, and pulse sampling resolutions.

Because Vivaldi antenna has a simple structure, lightweight, small lateral dimensions, and constant gain and radiation patterns, the UWB see-through-wall system uses 16 Vivaldi subarrays to make a full linear array, which allows for synthetic steering at the azimuth plane.

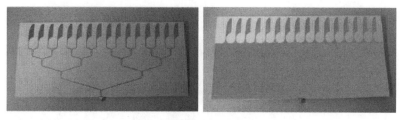

Figure 3.27 Subarray of the UWB see-through-wall radar antenna that covers 7–13 GHz (Yang and Fathy, 2009). Courtesy of Professor Aly Fathy, The University of Tennessee, Knoxville, TN.

Figure 3.27 shows the subarray that covers 7–13 GHz with a size of 7×16 in^2. The subarray has a constant gain of 13–14 dB and constant radiation patterns over the entire frequency range, which is desirable for UWB see-through-wall applications. Details of the design and characterization of the array can be found in Stockbroeckx and Vorst (2000), Greenberg et al. 2003, and Yang et al. 2008). To automate the linear array sequential operation, a single-pole 16-throw (SP16T) switch based on 15 MMIC single-pole double-throw switches in a binary tree structure was implemented. The switch driver is controlled by a 4-bit transistor–transistor logic TTL word corresponding to 16 logic states that turn on each of the 16 array channels. The SP16T switch operates from 1 to 12 GHz. The design complexity can be significantly reduced using off-the-shelf SP4T switches available on the market.

As the optimal waveforms for see-through-wall detection is still under study, the system adopts a tunable pulse generator that can generate both Gaussian and Monocycle waveforms. A step-recovery diode and an improved matching network were used to prevent the pulse echoing and minimize pulse width broadening. The pulse can be generated with an adjustable pulse-width from 300 ps to 1 ns. The details of the pulse generator design can be found in Zhang and Fathy (2006). Other front-end components include PAs, LNAs, mixers, power dividers, and hybrid couplers.

Experiments were carried out in the hallway of an office building. The microwave front-end and the FPGA imaging module were tested in the experiments. The see-through-wall experiment setup is shown in Fig. 3.28, where a panel is present between the Vivaldi array and the subject under test.

Figure 3.28 See-through-wall experiment set up by Yang et al. at the University of Tennessee (Yang and Fathy, 2009). Courtesy of Professor Aly Fathy, The University of Tennessee, Knoxville, TN.

In experiment, one of the 16 Vivaldi subarrays was used as the transmitting antenna while the rest of the subarrays formed a linear receiving array. The trace of a moving target was displayed on a computer monitor in real time. Figure 3.29 shows the radar images for a target moving at the maximum speed of 1.5 m/s. The radar displays responses within a 15 m distance range and a $-30°/30°$ azimuth-scanning range at a $1°$ angular resolution and 208 ps sampling resolution. The experiment successfully demonstrated a practical see-through-wall UWB radar system using efficient fast image processing. It is interesting to note that the same UWB radar system can also detect (through wall) the breath of a human body standing still up to 2 m in front of the radar. This reveals another approach for time/frequency domain signal analysis (instead of using 2D/3D image) for UWB see-through-wall detection. Moreover, by dividing the 16-element Vivaldi array to subarrays that are spatially distributed, a MIMO radar system with finer range resolution can be realized.

3.1.5 FMCW Radar

In recent years, the FMCW radar has been applied to motion sensing owing to its advantages of being able to operate with low power, and being able to detect both relative motion and absolute range. The

Figure 3.29 Real-time traces of moving target in see-through-wall experiment (Yang and Fathy, 2009). Courtesy of Professor Aly Fathy, The University of Tennessee, Knoxville, TN.

Systems Micro Technologies has built a micro-power FMCW radar prototype (Mostov et al., 2010), which contains radar antennas, front-end components, and initial signal processing units. Subsequent signal processing is performed on a computer. The prototype has a size of $15 \times 10 \times 5$ cm^3. The antenna has a bandwidth of 10–11 GHz. The radar measurement range is 0.5–10 m with a range resolution of 0.5 mm and radiated power of 0.8 mW. The range resolution is defined as the detectable change in the target range. The target resolution, which is defined as the minimum resolvable distance between individual targets, is 15 cm. The radar's field of view is 50°, and its data sampling rate is 120 Hz.

As shown in Fig. 3.30, three sets of human subject experiments have been conducted by Mostov et al. using the micro-power FMCW radar to validate its ability to measure heart rate and monitor respiration.

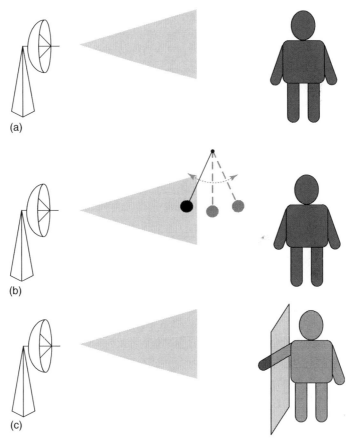

Figure 3.30   Three sets of experiment to test the FMCW radar. (a) Measuring breathing and cardiac rates of a human subject. (b) Measuring two targets simultaneously: a small ceramic pendulum was placed halfway between the radar and the human subject. (c) Measuring respiration from part of a human body. Adapted from Mostov et al., (2010).

The measurement data were collected for 1 min when the target is 2 m away from the radar.

In the first experiment of Fig. 3.30a, the radar beam was projected to the subject's chest. A standard pulse oximeter and a Vernier respiration monitor belt were used to provide heart and breathing rates respectively as the reference. The radar measured results were compared with the reference to evaluate the radar's accuracy. During the 1-min measurement, both radar and the reference methods provided the same heart rate of 78 bpm (beats per minute) and breathing rate of 15 bpm. In the second experiment of Fig. 3.30b, the previous experiment was modified to validate the radar's ability to measure more than one target by adding a pendulum at 1 m range from the radar antennas. The pendulum, made from a 2-cm-wide and 5-cm-long ceramic cylinder suspended on a string, swung at 27 bpm. Because of the FMCW radar's range detection ability, analysis was performed separately for different values of range (1 and 2 m). As a result, both the pendulum motion frequency and the vital sign signals (heartbeat and breathing) were successfully recovered. In the third experiment of Fig. 3.30c, the radar's sensitivity to weak signals was tested. The subject's torso, head, and legs were screened from the radar by a conducting copper mesh with 1 mm spacing so that only a forearm and a hand were "visible" to the radar antennas. The subject was instructed to breathe following a specific pattern: five 5-s breaths, one 18-s breath-hold, and then six 3-s breaths. The breathing pattern was verified using the Vernier respiration monitor belt. Both the air-pressure belt and the radar signals exhibit the same time dependence up to a scaling factor, confirming that the radar can be used to follow respiratory activity in real time even if the direct view of patient's chest is obstructed.

The existing works of using FMCW radar for motion detection has demonstrated FMCW radar's advantages of both high detection sensitivity and good range resolution. It is a good candidate for many portable low power applications.

3.2 RADAR TRANSPONDERS

A transponder is a device that emits an identifying signal in response to a received interrogating signal. In radar motion sensing applications, the transponders help identify the signal reflected from the target while rejecting noise and interferences from surrounding environment.

3.2.1 Passive Harmonic Tag

Harmonic tags can be used to isolate the motion of desired targets from the motion of untagged clutter in radar motion systems. Therefore it can improve the SNR of motion sensing radar. A typical harmonic tag consists of a tag antenna with a strongly nonlinear element connected to the antenna port. In the works of Singh et al. (2009), the nonlinear element is a Schottky barrier diode. The radar sends out a signal to the harmonic tag, where the nonlinear Schottky barrier diode up-converts the incoming signal into harmonic frequencies. The passive harmonic tag is designed so that the second harmonic is transmitted back to the receiver.

One application of the passive harmonic tag is the noncontact vital sign detection. Any object moving nearby the person under test will cause interference to the radar. The motion at other parts of the body (e.g., abdominal movement and arm motion) will also affect the accuracy of respiration and heartbeat detection. Adding filters does not solve the problem due to the random behavior of the interference sources. Singh et al. (2012) placed harmonic tag on the subject's chest so that only the tag motion is detected and other interference motion in proximity to the subject is rejected by the receiver to a reasonable extent. Figure 3.31 shows the setup of the system with harmonic tag attached to one person to reject the interference caused by another person.

To optimize the efficiency, the antenna in the passive harmonic tag should be designed to present appropriate impedance to the nonlinear device. A simple approach is to use a wire dipole with the Schottky diode attached across an inductive loop in the dipole. However, it could not be applied to the radar motion sensing application because the systems normally require a low transmitting power. In order to improve the performance with a low power level, the planar tag was designed with a copper tape, as shown in Fig. 3.32. A double dipole structure with the dipoles inductively loading each other was utilized to match with the Schottky barrier diode. The tag impedance was chosen to be close to the conjugate of the diode reactance at both 2.45 GHz and its second harmonic of 4.9 GHz. The tag was implemented on a 0.5 cm Styrofoam substrate in order to minimize the effect of the human body on the EM field. For a 10 dBm transmitted power at 2.45 GHz from the radar, the received signal from the tag at

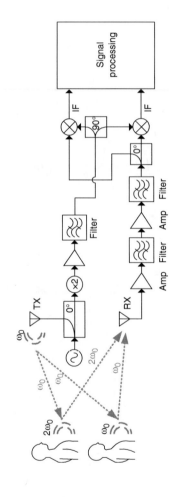

Figure 3.31 A radar noncontact respiratory monitoring system with passive harmonic tags. Adapted from Singh et al. (2012).

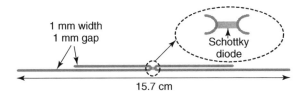

Figure 3.32 The planar tag used in the passive harmonic tag system. Adapted from Singh et al. (2012).

4.9 GHz was measured as −87 dBm at a detection distance of approximately 0.7 m. The radar transmitted power level has to be chosen so that it does not affect the tag activation range in typical close range monitoring applications.

3.2.2 Active Transponder for Displacement Monitoring

Because a passive harmonic tag needs high power to be turned on and it introduces significant power loss, harmonic radar has limited range performance of less than a few meters. Therefore, passive harmonic tags are not appropriate when a radar displacement sensor is used for applications with larger detection range (e.g., monitoring the structural response of an overpass bridge). Moreover, harmonic radar is difficult to control because a passive tag reflects the second harmonics whenever it receives RF power, which has limitations in sensor network applications.

Radar sensor system with active transponders can be used to solve the above difficulties. Active transponders convert the radar-transmitted signal to a different frequency, which is received by the radar receiver. In this way, only the tagged target is monitored and the backscatter from clutters at the radar-transmitted frequency causes very little interference. Moreover, the range performance of the radar can be improved because active transponders can amplify the signal level when sending it back.

Figure 3.33 shows an example pair of radar sensor and active transponder. Two 2.4/3.3 GHz patch antennas and two monopole 900 MHz antennas were custom designed for system evaluation. The radar sensor, shown in the inset of Fig. 3.33, was designed with a coherent heterodyne architecture. It transmits a single-tone 2.4 GHz frequency to the active transponder. In the transponder, an LNA and

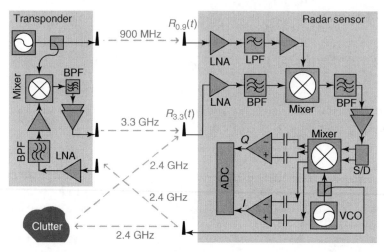

Figure 3.33 Example of a radar sensor with an active transponder. From Gu et al. (2012d).

a gain block boost the signal to drive the passive double-balanced mixer. A 2.4 GHz bandpass filter blocks the out-of-band tones. The 2.4 GHz signal is then up-converted to 3.3 GHz by mixing with a 900 MHz signal from a signal generator in the transponder. The 3.3 GHz signal is filtered and then amplified by two stages of gain blocks before it is sent back to the radar sensor. The signal received at the 3.3 GHz RF front-end of the radar sensor is

$$R_{3.3}(t) = A_{3.3} \cos\left(2\pi\left(f_1 + f_2\right)t - \frac{2\pi x(t)}{\lambda_1} - \frac{2\pi x(t)}{\lambda_3} + \Delta\varphi\right),$$
(3.2)

where $A_{3.3}$ is the amplitude of the received 3.3 GHz signal, $f_1 = 2.4$ GHz, $f_2 = 900$ MHz, λ_1 is the wavelength of the 2.4 GHz signal, λ_3 is the wavelength of $f_3 = f_1 + f_2 = 3.3$ GHz signal, $x(t)$ is the target motion, and $\Delta\varphi$ is the residual phase noise including the phase offset determined by the initial target position. The 900 MHz signal received at the radar sensor also has phase modulation due to the relative motion between the transponder and the radar sensor:

$$R_{0.9}(t) = A_{0.9} \cos\left(2\pi f_2 t - \frac{2\pi x(t)}{\lambda_2} + \Delta\theta\right),$$
(3.3)

where $A_{0.9}$ represents the amplitude of the 900 MHz signal, λ_2 is the wavelength of the 900 MHz signal, and $\Delta\theta$ is the residual phase noise. The received 3.3 GHz signal is mixed with $R_{0.9}(t)$ to produce a 2.4 GHz IF that is amplified and further down-converted to baseband I/Q signals by mixing with the same 2.4 GHz transmitted signal. The baseband I/Q signals output is

$$I \text{ channel: } A_{\mathrm{BB}} \cos\left(2\pi x\,(t)\left(\frac{1}{\lambda_1} + \frac{1}{\lambda_3} - \frac{1}{\lambda_2}\right) + \Delta\delta\right), \quad (3.4)$$

$$Q \text{ channel: } A_{\mathrm{BB}} \sin\left(2\pi x\,(t)\left(\frac{1}{\lambda_1} + \frac{1}{\lambda_3} - \frac{1}{\lambda_2}\right) + \Delta\delta\right), \quad (3.5)$$

where A_{BB} and $\Delta\delta$ are the amplitude and residual phase noise. The target motion $x(t)$ can be recovered using a microwave interferometry technique.

The range performance of the proposed radar sensor system was tested in lab environment and its ability to reject clutters was analyzed. In experiment, the radar sensor was placed on a linear actuator (Zaber TSB60-I) that mimics the vibration of a building structure, and the active transponder was placed at a stationary point a few meters away from the radar. During the experiment, the movement of the linear actuator was detected by the radar sensor with the active transponder as a reference.

To evaluate the accuracy of radar sensor with active transponder, the actuator was first programmed to vibrate with a frequency of 0.25 Hz and amplitude of 3 mm. The sinusoidal motion was recovered from the radar I/Q outputs. It was then evaluated in the frequency domain to compare its signal strength and frequency with the programmed 0.25 Hz 3 mm movement. The radar measured signal matches very well with the programmed movement in both amplitude and frequency. Second, the actuator was programmed to move linearly at the speed of 2.6 mm/s. The recovered linear motion was also compared with the programmed movement. The results show that the radar accurately tracked the linear motion with a maximum displacement error of less than 1.0 mm. The SNR and detection error of the proposed radar system was evaluated and summarized in Table 3.4.

To demonstrate clutter noise rejection using active transponder, the radar sensor was used to measure vibration motion when

TABLE 3.4 Summary of Radar Measurement Result (Gu et al., 2012d)

	Sinusoidal			Linear
	Amplitude, mm	Frequency, Hz	SNR, dB	Error, mm
Measured	2.96	0.25	47.623	<1.0
Programmed	3.00	0.25	N/A	N/A

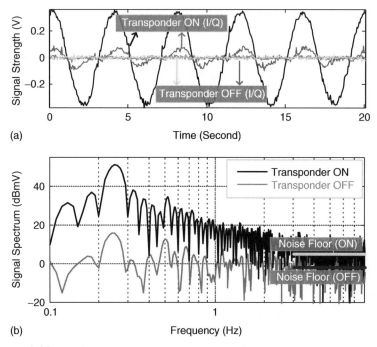

(a)

(b)

Figure 3.34 Radar output when measuring 0.25 Hz vibration at 4.1 m away. (a) Time domain I/Q signals when the transponder was turned ON/OFF. (b) Corresponding spectra.

the transponder was turned ON/OFF. The responses are shown in Fig. 3.34. It is shown that the transponder boosted the signal strength by over 35 dB. The noise floor was calculated by averaging the spectrum from 3 to 8 Hz where the background white noise dominates. When the transponder was turned on, the measured signal was 50.89 dBmV on a 5.12 dBmV noise floor (SNR = 35.77 dB). However, when the transponder was turned off, the signal dropped to 15.40 dBmV and noise floor became −1.7 dBmV (SNR = 17.1 dB). Considering 15.40 dBmV as undesired background noise mainly due to leakage of the reflected 2.4 GHz signal into the sensor receiver, the system has an SNR of over 35 dB when detecting from 4.1 m away. This gives sufficient margin to further extend the measurement range beyond 4.1 m. As the transponder is small and only boost signals targeting at it, it allows accurate measurement of the vibration motion between fixed points.

3.3 ANTENNA SYSTEMS

Antenna is an indispensable part of a noncontact motion sensing system. It strongly affects the performance such as the detection accuracy, sensitivity, and space resolution. As early as the works performed at Michigan State University in the 1980s (Chen et al., 1986), several types of antenna systems have been designed and compared for motion sensing applications. They include reflector antenna, patch antenna, and probe antenna. Implemented in microwave life-detection system, they demonstrated the capability to locate human subjects buried under earthquake rubble or hidden behind various barriers by sensing the physiological motion caused by breathing and heartbeat. As then, more antenna types and configurations have been applied to radar noncontact sensing in different environments. This section will discuss phased array antenna, broadband microstrip antenna, and helical antenna.

3.3.1 Phased Array Systems

Phased arrays have been widely applied for military and commercial purposes for its advantages of high gain, focused beam, and thus superior spatial resolution in detection. In applications that the target is moving or multiple targets have to be measured, beam-scanning phased array greatly extends the capability of radar motion sensor. By electronically controlling the responses of each antenna, the radiation beam of radar system can be arbitrarily changed, resulting in fast beam scanning over the space.

The principle of the beam-scanning of phased array is briefly reviewed here. The array factor of a linear phased array of N element is given by

$$AF(\theta) = \sum_{n=1}^{N} e^{j(n-1)\varphi}, \qquad (3.6)$$

where $\varphi = k_0 d \cos\theta + \Phi$, k_0 is the free-space wavenumber, Φ is the element to element phase shift, d is the distance between adjacent elements, and θ is the angle of the radiation beam. The major radiation beam of the array is achieved when $\varphi = 0$. Under this condition, the radiation beam angle is related to Φ, which is expressed as

$$\theta = \arccos\left(\frac{\phi c}{2\pi f d}\right), \quad \text{since } \phi = -k_0 d \cos\theta = -\frac{2\pi f}{c} d \cos\theta,$$
$$(3.7)$$

where f is the operating frequency, and c is the speed of light. According to Equation 3.7, if Φ (the phase difference between adjacent elements within the array) is changed, the radiation beam of the array will be changed. Therefore, beam scanning can be realized by controlling the value of Φ.

Compared with the conventional radar with simple antenna, there are two advantages offered by beam-scanning radar. First, the scanning of detection beam allows less radar to be needed to fulfill certain requirements. Second, as the phased array used for beam-scanning usually feature higher antenna gain, it will improve the SNR of the detection. This can further enhance the measurement/tracking resolution, leading to more accurate measurement.

3.3.2 Broadband Antenna

As the optimal carrier frequency of motion sensor varies with different motion characteristics, in some applications it is desirable to have a broadband antenna so that the radar can be dynamically tuned to different operating frequencies.

There have been many design approaches for planar wideband antennas using partial ground planes. A wideband patch antenna consisting of a rectangular patch with two steps, a single slot on the patch, and a partial ground plane was reported in Choi et al. (2004). A further modification of partial ground plane patch antenna was introduced by adding a bent stub on the radiating patch and stepping the ground plane. By inserting an inverted U-shaped slot on the circular patch, the antenna showed a band-rejection characteristic (Choi et al., 2005). The effects of ground plane dimensions on antenna performance such as gain, bandwidth, and radiation pattern were investigated in Curto et al. (2007). These wideband patch antennas operated approximately from 3 to 12 GHz.

In addition, there have been design approaches using parasitic patches to control the peak directivity or improve the bandwidth of microstrip antennas. The Yagi-antenna concept was introduced to control the peak directivity. With parasitic director and reflector patches located on the same plane of the driven element, the peak directivity, by the effect of mutual coupling, was tilted toward the end fire direction with peak gain of 8 dBi (Huang, 1989). The antenna was later used in an array for mobile satellite applications (Huang and Densmore, 1991). To improve the bandwidth, parasitic patches on all four edges including nonradiating edges were used to increase the impedance matching bandwidth to 815 MHz (Kumar and Gupta, 1985).

Several broadband microstrip antennas have been developed for motion sensing applications. In Park et al. (2009), a design combining a partial ground plane and parasitic patches is presented. The partial ground plane improves broadband characteristics and the parasitic patches improve the gain. As a result, a 4.3–7 GHz broadband antenna with improved gain was achieved, as shown in Fig. 3.35. Compared with a partial ground plane single patch reference antenna, the designed broadband antenna shows better directivity (narrower beam patterns toward intended directions) and improved gain. It also shows improved isolation between the transmitting and receiving antennas when implemented in a 4.4–6.7 GHz vital sign detection radar. The radiation pattern was designed for mounting on the ceiling at the corner of a room or a hallway.

The radiation patterns of this antenna and a partial ground plane single patch reference antenna are shown in Fig. 3.36. In particular, E-plane copolarization and cross-polarization radiation patterns were measured at frequencies of 4.5, 5.5, and 6.5 GHz, and compared with simulations. It was shown that the peak directivity was tilted by about 35° from the broadside toward the end fire direction, which is caused

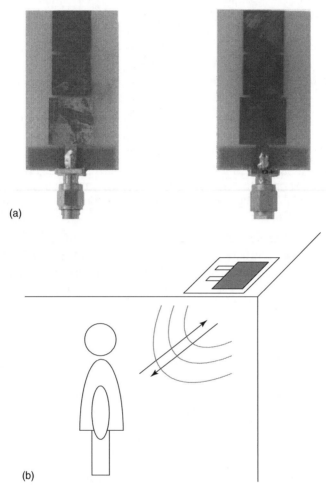

Figure 3.35 (a) The front and the back of a 4.3–7 GHz microstrip antenna for indoor noncontact vital sign detection. (b) Illustration of the radar installation (Park et al., 2009).

by the effect of mutual coupling. A $35°$ of peak directivity angle is chosen for installation on the ceiling at a corner location. If the angle is tilted too close to the end-fire direction, the beam will be almost parallel to the ceiling. If the angle is close to 0, that is, broadside direction, the radar detection will be limited to a spot just below where the

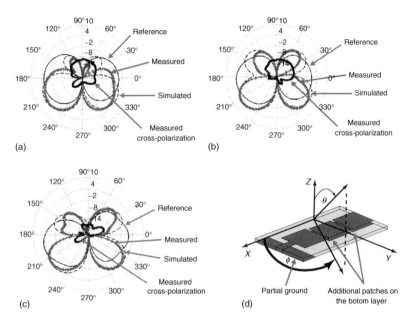

Figure 3.36 Comparison of simulated/measured radiation patterns of the proposed antenna and the reference antenna along $\Phi = 0$ plane (Solid: Reference, Dashed: Proposed antenna—simulated, "++": Proposed antenna—measured, "xx": Proposed antenna—measured cross polarization). Radiation patterns at frequencies of 4.5 GHz (a), 5.5 GHz (b), and 6.5 GHz (c) are plotted. Antenna configuration with axes is shown in (d) (Park et al., 2009).

antenna is installed. Neither case is desired for the intended application. The tilted angle is a function of dielectric constant, substrate thickness, patch separation, gap distance, and sizes of the parasitic patches. Therefore, the tilted angle can be modified by varying the above parameters and the number of patches.

3.3.3 Helical Antenna

In some applications, it is desirable to have a narrow antenna beamwidth so that the detection can be focused on the target and reject the noise from clutter. Helical antenna is a good candidate to achieve such a goal. In Fletcher and Han (2009), helical antennas were

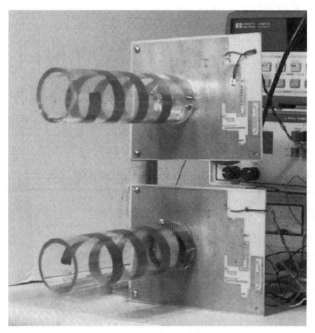

Figure 3.37   Helical antenna pair operating at the 2.4 GHz frequency band. (Fletcher and Han, 2009). Courtesy of Dr. Richard Fletcher, MIT Media Lab, Cambridge, MA.

constructed by winding copper tape on a low loss acrylic tube. The circumference C of the tube was chosen to match the wavelength λ. The length of the tube, which is determined by the number of turns, was chosen to meet the desired beamwidth requirements. The pitch of the windings was experimentally determined by optimizing the antenna gain and specified by the turn spacing S, such that the pitch equals $\arctan(S/\pi D)$. Both left-handed and right-handed polarization versions were fabricated. Figure 3.37 shows the helical antenna designed for a Doppler radar system operating at the 2.4 GHz frequency band. The specifications of the antenna are: frequency = 2.45–2.5 GHz, wavelength $\lambda = 12.2$ cm, turn space = 0.27λ, number of turns = 4, circumstance $C = 1.14\lambda$. The antenna achieves a half-power beamwidth of 42°.

Utilizing this helical antenna pair, a differential front-end was realized to improve the performance of short-range low cost Doppler radars for vital sign detection with applications to automotive driver safety systems, health monitoring, and security screening. The dual helical antennas illuminate the body in two adjacent locations to perform a differential measurement. As only one of the beams illuminates the heart, the baseband signal from the second radar is used for motion cancellation. In experiment, reduction in background motion noise as compared with traditional single radar unit design was achieved. Details of the differential front-end will be discussed in Section 4.1.

4

ADVANCES IN DETECTION AND ANALYSIS TECHNIQUES

4.1 SYSTEM DESIGN AND OPTIMIZATION

Several advanced detection techniques to improve the performance of motion radar sensors used for vital sign and vibration detection are introduced in this section. These techniques involve novel system architectures and detection algorithms. A major focus in recent development effort is to eliminate the "clutter" noise and interference caused by unwanted reflections from other objects surrounding the target, or the random motion artifacts caused by the target itself. In addition, accurate measurement of not only the rate but also the displacement of a sinusoidal or nonsinusoidal vibration is explored.

4.1.1 Shaking Noise Cancellation Using Sensor Node Technique

As the phase stability of the radar measurement system plays an important role in successful detection of motion signals, small unwanted mechanical motions of the radar transmitting antenna cause unrecoverable phase errors in the received signals. This problem severely affects

Microwave Noncontact Motion Sensing and Analysis, First Edition.
Changzhi Li and Jenshan Lin.
© 2014 John Wiley & Sons, Inc. Published 2014 by John Wiley & Sons, Inc.

the radar performance especially in the case when the radar is a mobile or handheld device. To address this issue, Mostafanezhad et al. (2007) proposed to use a bistatic radar with a sensor node receiver placed in the vicinity of the subject. The sensor node consists of an antenna and a mixer. It receives both the direct signal from the transmitter LO and the signal reflected from a human subject. Both signals are subject to the same "mechanical" phase noise caused by unwanted motion of the radar transmit antenna. If these path lengths are similar, there will be significant phase noise reduction due to the range correlation effect, making accurate detection of motion signals possible.

An example setup using a bistatic radar with a receiver (i.e., sensor node) placed in the vicinity of the subject is shown in Fig. 4.1. The signal transmitted from an ideal CW radar can be represented as

$$S_t(t) = \cos(\omega_0 t), \tag{4.1}$$

where ω_0 is the frequency of LO. The signal reflected from the subject will be received by the radar receiver as

$$S_r(t) = A \cos\left(\omega_0 t - \frac{4\pi R_{tb}}{\lambda}\right), \tag{4.2}$$

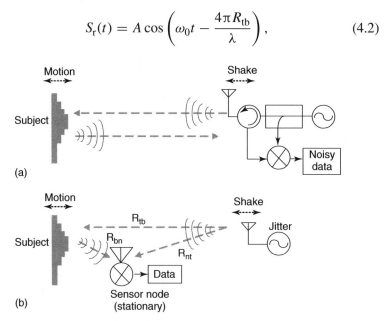

Figure 4.1 Shaking noise cancellation using sensor node technique. (a) A monostatic radar; (b) a sensor node configuration. Adapted from Mostafanezhad et al., (2007).

where λ is the wavelength and R_{tb} is the time-varying distance between the subject and the radar antenna. Demodulating the received signal using the radar LO signal will obtain the information of R_{tb}, but it will be inevitably affected by the noise caused by radar shaking. On the other hand, the total RF signal received at the sensor node antenna is

$$S_{nRF}(t) = B \cos\left(\omega_0 t - \frac{\omega_0}{c} R_{nt}\right) + C \cos\left(\omega_0 t - \frac{\omega_0}{c} R_{tb} - \frac{\omega_0}{c} R_{bn}\right),$$
(4.3)

where R_{tn} is the time-varying distance of radar transmitter to the sensor node and R_{bn} is the time-varying distance from the subject to the sensor node. If the amplitude variation due to propagation loss is neglected, mixing $S_{nRF}(t)$ with itself by passing it through a nonlinear device will result in the following baseband component at the sensor node output

$$S_n(t) = BC \cos\left(\frac{2\pi}{\lambda}\left(R_{tb} + R_{bn} - R_{nt}\right)\right).$$
(4.4)

If the monostatic antenna is located at a large distance from both the human subject and the node, such that $R_{tb} \approx R_{nt}$, slight physical movements of the monostatic antenna have the same effect on R_{tb} and R_{nt}, so that they cancel each other and result in the following baseband output at the sensor node

$$S_n(t) \approx BC \cos\left(\frac{2\pi}{\lambda} R_{bn}\right).$$
(4.5)

The above analysis illustrates that, compared with the monostatic radar, the received signal at the sensor node is less sensitive to the unwanted movements of the monostatic radar antenna. This effect is similar to the range correlation effect, which reduces the baseband noise caused by the LO's phase noise. The two signals arriving at the sensor node contain nearly the same phase variation caused by unwanted movements of the monostatic antenna. The closer the node and the subject are, the better these two phase variations cancel out, resulting in a less noisy baseband signal and more accurate motion detection results.

Figure 4.2 shows the simulated outputs of the radar node and monostatic radar, for a subject displacement of 5 mm. The sensor node

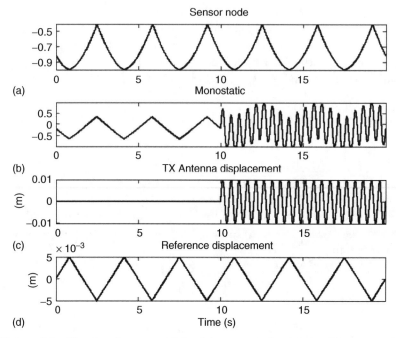

Figure 4.2 Simulated outputs of the (a) sensor node and the (b) monostatic radar for a subject with a displacement of 5 mm. (c) The monostatic antenna begins to shake with amplitude of 10 mm after 10 s. (d) While the sensor node output remains unchanged, the antenna shaking greatly affects the monostatic radar output. Copyright © IEEE. All rights reserved. Reprinted with permission from Mostafanezhad Park B, Boric-Lubecke O, Lubecke V, Host-Madsen A. Sensor nodes for Doppler rader measurements of life signs. IEEE MTTS-S Int Microw Symp Dig Honduly (HI) 2007:1241–1244.

is placed at 20 cm from the subject. The monostatic transceiver is located 2 m away from the subject and begins to shake with a 10 mm displacement after 10 s. During the first 10 s of the simulation interval, both the monostatic radar and the sensor node track the subject motion. Once the transmitter antenna begins to shake, this shaking is clearly reflected at the monostatic radar output, while the sensor node output is not affected and it continues to accurately track the subject motion. Experimental results have also been demonstrated in Mostafanezhad et al. (2007) to verify the theory and simulation.

4.1.2 DC-Coupled Displacement Radar

For displacement and vibration measurement, the radar suffers from DC offset caused by reflections from stationary objects around the target, which may saturate the following stages of baseband amplifiers. This type of DC offset is inevitable for both direct-conversion and indirect-conversion architectures and is difficult to deal with, as it depends on different test environments. To overcome this problem, AC coupling has been commonly used in vibration radar sensors. However, AC coupling causes significant signal distortion when the target motion has a very low frequency or a DC component, due to the high pass characteristics of the coupling capacitor. This is a problem in Doppler radar motion sensing when a target has stationary moment. To deal with it, researchers have proposed several approaches to employ DC coupling in radar sensors. Park et al. (2007) proposed a method for calibrating the DC offset while preserving the DC information. However, extra effort is demanded before real-time measurement. In Zhao et al. (2011), low DC offset was achieved using mixers with high LO-RF isolation. However, this method only alleviates the DC offset from circuit imperfection, and the remaining DC offset due to reflection from stationary objects still limits the dynamic range of the baseband amplifiers. To eliminate the DC offset due to reflections from stationary objects, an adaptive DC-coupled radar sensor using fine-tuning adaptive feedback loop was proposed (Gu and Li, 2012c). The key technology in this radar is a real-time feedback loop, which extracts the DC offset information and adaptively adjusts bias to a level that allows high gain amplification at the baseband stage.

Figure 4.3 shows a prototype of the adaptive DC-coupled radar sensor with direct-conversion quadrature architecture. The down-conversion mixer of the radar receiver is DC coupled to the baseband amplifier. The I/Q channel baseband outputs are digitized by a DAQ (NI-USB6009) that connects to a laptop via a universal serial bus (USB) port. Through the general purpose interface bus (GPIB), the laptop controls a power supply (Tektronics PS2521G) that is able to sweep over a wide range of voltages. Unlike the conventional AC-coupled radar sensor, the proposed DC-coupled radar sensor allows the external power supply to adjust the biasing level at the baseband amplifier. Take the Q channel, for example, assuming that the amplifier

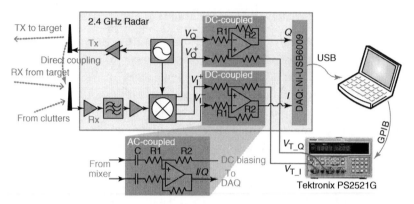

Figure 4.3 Block diagram of the prototype adaptive DC-coupled radar (Gu and Li, 2012c).

has infinite open-loop gain, the DC level of the amplifier output is

$$V_{Q\_out} = V_{T\_Q} + \left(\frac{R_2}{R_1}\right) \cdot (V_Q^+ - V_Q^-),$$

where V_{T_Q} is the tuning voltage from the power supply, V_Q^+ and V_Q^- are the DC offsets of the differential channels, and R_2/R_1 is the close-loop gain. Owing to the reflections from stationary objects and circuit imperfection, the DC difference $(V_Q^+ - V_Q^-)$ is not equal to zero. Therefore, with a desirable high baseband gain, even a small difference between the incoming differential signals would be significantly amplified, making the second term in the Q channel output equation very large. If V_{T_Q} is fixed at a certain point, the amplified DC offset could easily saturate the amplifier. For example, if the amplifier power supply is 3.3 V, R_2/R_1 is set to be 180, $(V_Q^+ - V_Q^-)$ DC component is 0.04 V, and V_{T_Q} is fixed at half of the power supply voltage (1.65 V), the V_{Q_out} DC component reaches at 7.2 V, which makes the Q channel output to clamp at 3.3 V. For the prototype adaptive DC-coupled radar sensor, the output V_{Q_out} is analyzed by the laptop in real time. Depending on the DC value of V_{Q_out}, the laptop configures the power supply to decrease V_{T_Q} when the output is close to the lower rail, or increase V_{T_Q} when the output is close to the upper rail. This process continues until the amplifier output reaches the desired DC level that is close to the middle between the supply rails. This allows sufficient high gain and avoids saturation. Thus, an adaptive feedback loop is formed. Because the tuned DC level can be easily compensated back in software before further signal processing, the DC information is not lost in this solution. The fine-tuning feature also allows the radar sensor to work with the largest dynamic range.

Experiments were performed to demonstrate effective operation of the adaptive DC-coupled radar. In the experiments, a motion phantom from Varian Medical Systems was used as the target. The phantom exhibits sinusoidal-like 0.5 Hz motion but with a short stationary period at the end of a motion cycle. Three radar sensors with different baseband circuit structures were used in the experiments: (i) DC coupled, (ii) 10 μF AC coupled, and (iii) 30 μF AC coupled. The radar sensors detected the phantom motion from about 0.5 m away. A camera was placed 1 m away from the phantom to record the motion of the phantom. The recorded video was processed to extract the phantom motion, which was then used as a reference to compare with the signal measured by the radar sensors. A software interface developed in C# was applied to communicate via GPIB with the power supply and controls its sweep steps and intervals.

The adaptive feedback loop was first tested for its ability to fine tune the DC offset level. V_{T_Q} and V_{T_I} were originally set to be 1.65 V. After reading the I/Q outputs, the laptop sets a negative step to tune the upper curve down and a positive step to tune the lower curve up. The tuning process continues until the amplifier output was tuned to the desired level. To verify the adaptive DC-coupled sensor's ability to relieve from saturation, the radar sensor measured the phantom motion for both cases with and without DC tuning. The results are shown in Fig. 4.4. In Fig. 4.4a, after DC tuning with steps of 0.5 V/0.8 V respectively, both channels were successfully adjusted step-by-step to a reasonable level. Fig. 4.4b shows DC tuning with finer step size of 0.05 V, which allows precise tuning to half of the amplifier power supply voltage. In real application, the tuning procedure can be made much faster than mechanical movements. With the phantom turned on, Fig. 4.4c shows the radar measurement results before and after a fast DC tuning. It is shown that the amplifier was originally affected by large DC offset, such that the I channel was saturated on the top and the Q channel was totally clamped on the bottom. However, after DC tuning, both channels exhibit satisfactory amplification to signals. The I/Q channels have different patterns because of different I/Q response to the residual phase.

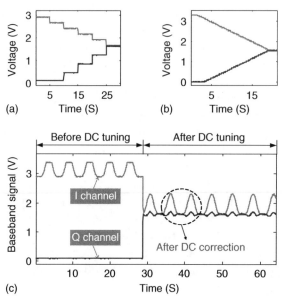

Figure 4.4   DC tuning process (a) and radar measurement result before after DC tuning (b).

In the second experiment, the camera and the three radar sensors were used to measure the same phantom motion. Figure 4.5 illustrates the results of the camera and three radar sensors measuring a pseudo-sinusoidal movement with stationary moment. After the I/Q signals were recorded, the microwave interferometry method was used to recover the displacement information. It is shown that the displacement measured by the proposed DC-coupled sensor matches very well with the reference displacement measured by the camera. On the other hand, both AC-coupled radar sensors suffered from distortion in the measured displacement. This is because during the stationary period, the capacitor cannot hold the charge for a long time and tends to charge or discharge. It is also demonstrated that larger capacitors do help alleviate the signal distortion due to lower cutoff frequency. However, larger capacitors also result in longer settling time, which may not be acceptable in some applications. The experiment demonstrated that the adaptive DC-coupled system is able to maintain DC information in the measured movement while the AC-coupled systems have signal distortion problem.

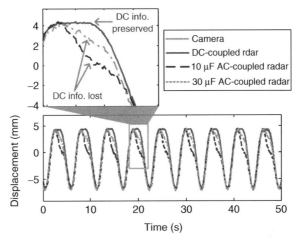

Figure 4.5 Movement measured by camera, adaptive DC-coupled radar sensor, and AC-coupled radar sensor.

Figure 4.6 shows another measurement result plotted in a constellation graph to compare the I/Q trajectories. It is seen that the trajectory of the DC-coupled radar matches well with the unit circle, while the AC-coupled radar tends to deviate from the circular arc to form a ribbon-like shape. This is due to the signal amplitude variation

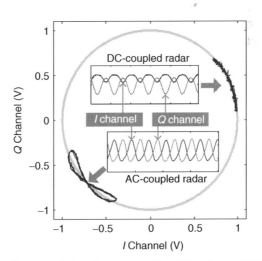

Figure 4.6 Trajectory of signals measured by AC radar and DC radar. Insets show the time-domain radar signals.

caused by the AC coupling capacitors' charging and discharging. As the DC-coupled radar has a trajectory that is much closer to an ideal arc, it leads to more accurate demodulation result of the original movement pattern.

4.1.3 Random Body Movement Cancellation Technique

Another problem caused by small unwanted mechanical motion is the random movement of a whole target when only part of the target's motion is of interest to the radar detection. A typical example is the radar noncontact vital sign detection scenario. The noise caused by the random body movement, which presents severe interference for accurate detection of respiration and heartbeat signals in practical applications, is one of the main challenges for noncontact vital sign detection. As random body movement is comparable or even stronger than the weak vital sign signal, to some extent, it is the main factor limiting broad applications of noncontact vital sensors. It will be shown in this section that the different patterns of random body movement and physiological movement make it possible to remove the unwanted signal from the vital signs. Typically, random body movement cancellation technique uses multiple antennas and transceivers to detect from the front and the back of a human body. Two solutions, that is, detection from different orientations and differential detection from one orientation, will be discussed in detail in this section.

4.1.3.1 *Detection from Multiple Radars at Different Observation Angles* On the basis of polarization and frequency multiplexing, signals detected from multiple radars located at different observation angles can be combined without interfering each other and noise caused by random body movement can be cancelled out. Following the spectral analysis in Chapter 2, for noncontact quadrature demodulation of vital sign, the signals detected by the I and the Q channels can be combined using the complex signal demodulation technique. The baseband complex signal output reconstructed in software can be expressed as follows

$$S(t) = I(t) + j \cdot Q(t)$$
$$= \exp\left\{ j \left[\frac{4\pi x_h(t)}{\lambda} + \frac{4\pi x_r(t)}{\lambda} + \phi \right] \right\}$$

$$= DC + 2j[C_{10}\sin(\omega_r t) + C_{01}\sin(\omega_h t) + \cdots] \cdot e^{j\phi}$$

$$+ 2[C_{20}\cos(2\omega_r t) + C_{02}\cos(2\omega_h t) + \cdots] \cdot e^{j\phi}, \quad (4.6)$$

where $x_h(t) = m_h \cdot \sin \omega_h t$, $x_r(t) = m_r \cdot \sin \omega_r t$ are the periodic body movements due to respiration and heartbeat, J_n is the Bessel function of the first kind, λ is the wavelength of wireless signal, Φ is the total residual phase accumulated in the circuit and along the transmission path. $C_{ij} = J_i(4\pi m_r/\lambda) \cdot J_j(4\pi m_h/\lambda)$ determines the amplitude of every frequency component. The DC components accumulated in the I and the Q channels only contribute to the DC term in the complex signal $S(t)$, which can be easily removed by subtracting the complex average from the time-domain sliding signal window before spectrum analysis.

As the chest wall movement caused by respiration and heartbeat is very small, random body movement presents a serious noise source for noncontact vital sign detection. Fortunately, the noise from random body movement can be eliminated by recognizing the movement patterns. Figure 4.7 shows the block diagram of the random movement cancellation technique. It should be noted that as the human body roams randomly in a certain direction (shown as the "Body" arrow in the figure), the heartbeat and the respiration cause the front and the back of the chest walls to move in the opposite directions. In the view of the two transceivers, the body movements due to heartbeat and respiration are in phase, while the random body movements are out of phase, that is, when the subject is roaming toward one transceiver, it is moving away from the other transceiver.

With the two-radar setup, the signal detected from the two transceivers can be expressed as

$$S_1(t) = \exp\left\{j\left[\frac{4\pi m_{r1}\sin(\omega_r t)}{\lambda_1} + \frac{4\pi m_{h1}\sin(\omega_h t)}{\lambda_1} + \frac{4\pi vt}{\lambda_1} + \phi_1\right]\right\}, \quad (4.7)$$

and

$$S_2(t) = \exp\left\{j\left[\frac{4\pi m_{r2}\sin(\omega_r t)}{\lambda_2} + \frac{4\pi m_{h2}\sin(\omega_h t)}{\lambda_2} - \frac{4\pi vt}{\lambda_2} + \phi_2\right]\right\}, \quad (4.8)$$

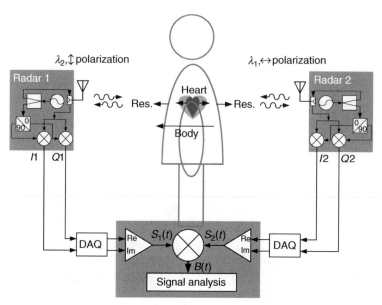

Figure 4.7   System setup of random body movement cancellation technique. Two transceivers, one in front of and the other behind the human body, are transmitting and receiving signals with different polarization and wavelength. The transceiver block diagram is simplified such that components not closely related to the algorithm are not shown (e.g., low noise amplifier and high gain blocks) (Li et al., 2008a).

where v is a random variable representing the velocity of random body movement. It should be noted that the desired physiological signal presents a *phase modulation* in the baseband signal, while the random body movement presents a random *frequency drift* in the baseband signal. By multiplying the two complex signals, the output $B(t)$ can be obtained as

$$B(t) = S_1(t) \cdot S_2(t)$$

$$= \exp\left\{j\left[4\pi\left(\frac{m_{r1}}{\lambda_1} + \frac{m_{r2}}{\lambda_2}\right)\sin(\omega_r t) + 4\pi\left(\frac{m_{h1}}{\lambda_1} + \frac{m_{h2}}{\lambda_2}\right)\sin(\omega_h t)\right.\right.$$

$$\left.\left. + \left(\frac{1}{\lambda_1} - \frac{1}{\lambda_2}\right)4\pi vt + \phi_1 + \phi_2\right]\right\}$$

$$\approx \exp \left\{ j \left[\frac{4\pi \left(m_{r1} + m_{r2}\right)}{\lambda} \sin(\omega_r t) + \frac{4\pi(m_{h1} + m_{h2})}{\lambda} \sin(\omega_h t) \right. \right.$$

$$\left. \left. + \ \phi_1 + \phi_2 \right] \right\}, \tag{4.9}$$

where the approximation is valid because λ_1 and λ_2 are chosen to be close to each other. The above operation corresponds to convolution and frequency shift in frequency domain, thus canceling the Doppler frequency drift and only keeping the periodic Doppler phase effects.

Simulations have been performed to verify this technique. The time-domain signals detected from the front and the back of the human body are generated by assuming the random body movement has a maximum of 5 cm displacement from the original body position as shown in the inset of Fig. 4.8a. The spectrum of the signal detected in each

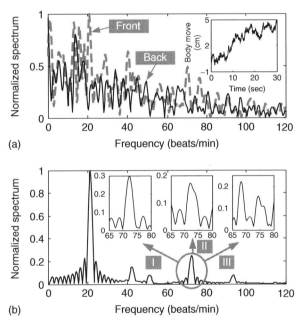

(a)

(b)

Figure 4.8 Simulation result of random body movement cancellation. (a) Spectrum detected from the front and the back of the subject, the body is moving randomly as shown in the inset. (b) Spectrum recovered by the random body movement cancellation technique when DC offset is calibrated out. Inset: heartbeat spectrum when DC offset/signal amplitude = 0.2 (I), 0.4 (II), and 0.6 (III).

channel and the spectrum of the complex signal are shown in Fig. 4.8. It is observed that the random body movement can be removed from the complex signal, and clear spectrum of desired signals can be obtained.

However, the random body movement cancellation technique is not immune from DC offset. The inset of Fig. 4.8b shows the spectrum around the heartbeat frequency when DC offset is present in the system. As the amplitude of DC offset increases from 20% to 60% of the desired signal amplitude, more noise is added to the spectrum until the desired signal is completely overwhelmed by noise.

As the two transceivers are facing each other, it is important to prevent signal of one unit from saturating or interfering the receiver link of the other unit. Therefore, patch array antennas with orthogonal polarization pattern are used for the two units. And free running VCOs are used for the two transmitters so that λ_1 and λ_2 are close to each other but always have a slight difference because the system does not incorporate any phase-locked-loop. Therefore, the signal from one transceiver can be easily rejected by the other transceiver in the baseband, because the small difference in the carrier frequency results in a large difference in baseband frequency for vital sign detection, which is typically no higher than several Hertz.

Experiments have been performed in lab environment to verify the random body movement cancellation techniques. The measurements were performed by 5–6 GHz portable radars, which integrate the quadrature transceiver, the two-stage baseband amplifier, and the power management circuit on a single Rogers printed circuit board (RO4350B) with a size of 6.8 cm × 7.5 cm. The amplified baseband output signals were sampled by a 12 bit multifunction DAQ and were fed into a laptop for real-time signal processing by LabVIEW. Figure 4.9 shows the antennas and the identical transceivers used for experiments.

Figure 4.10 shows a measurement result when two transceivers were used to verify the random body movement cancellation technique. During experiment, the subject under test was gently changing position in a chair, so that the noise of random body movement was emphasized. As the physiological movement at the back chest wall is weaker than that at the front chest wall, the noise completely overwhelmed the physiological signal from the back, and only overwhelmed the heartbeat signal from the front. When the proposed technique was applied

Figure 4.9 Two identical transceivers used in experiments. Inset: the antenna used for each transceiver. Note that one transceiver uses vertically polarized antenna array, while the other uses horizontally polarized antenna array.

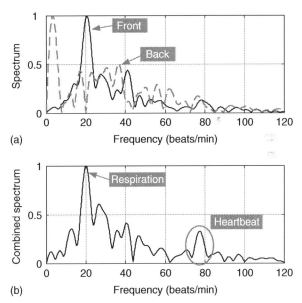

Figure 4.10 Detected baseband spectrum by the random body movement cancellation technique. (a) Complex signal demodulated spectrum measured from the front and the back of a human body. (b) Output spectrum by the random body movement cancellation technique, the heartbeat information is successfully recovered.

to combine the signals detected from the front and the back of the human body, it is shown in Fig. 4.10b that the heartbeat signal was recovered. Therefore, using multiple radars to detect from different sides of a human body is an effective way to cancel the noise produced by random body movement based on the different movement patterns presented.

4.1.3.2 Differential Detection   Another solution to improve the performance against random body motion is based on a differential Doppler radar front-end operating at two different frequencies. Fletcher and Han demonstrated that, by using dual helical antennas each with a 40° beamwidth, it is possible to illuminate the body in two adjacent locations to perform a differential measurement (Fletcher and Han, 2009). The experimental setup is shown in Fig. 4.11. The beam from the upper antenna illuminates the heart, while the beam from the lower antenna is used for motion compensation by detecting body motion.

A dual helical antenna and simple direct-conversion radar were implemented to demonstrate this approach. Although, the two radars all operate in the 2.4 GHz industrial, scientific, and medical (ISM) band, they were designed to have a relatively narrow bandwidth so that each of them operates with a separate frequency. Signal polarization discrimination between the two radars was also enabled based on the dual helical antenna. Although it is possible to combine the signal from each antenna to build a measurement matrix, the first demonstration of differential detection combined signals in baseband. The outputs of

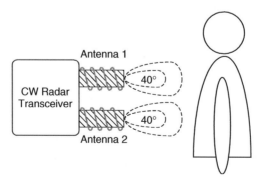

Figure 4.11   Dual-antenna Doppler radar for noise reduction in noncontact vital sign detection. Adapted from Fletcher and Han, (2009).

the two radar units were fed to an instrumentation amplifier to perform "baseline subtraction" before the signal was further processed. Active filters based on operational amplifiers were used for baseband signal filtering. The filters consist of two 6-pole elliptic filter sections with zeros at 60 Hz and corner frequencies of 80 and 100 Hz, respectively.

Figure 4.12 compares the physiological signal detected using a single radar with that obtained from differential detection using dual radar unit. Improvement in detection quality (i.e., reduced background motion noise) is demonstrated. However, as the motion of the chest

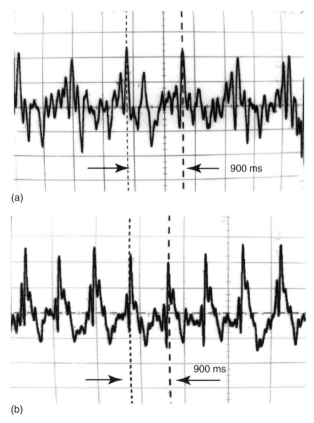

Figure 4.12 Physiological signal obtained by a single radar (a) and differential detection using dual-radar unit (b). Copyright © IEEE. All rights reserved. Reprinted with permission from Fletcher R, Han J. Low-cost differential front-end for Doppler rader Vital Sign monitoring. IEEE MTTS-S Int Microw Symp Dig 2007:1325–1328.

and diaphragm are not exactly equivalent within the beam spot zone of each radar (Fletcher and Han, 2009), further development of advanced signal processing is necessary to realize robust performance and complete removal of motion noise. Nevertheless, the differential detection in this approach provides significant performance improvement and enables further signal processing by reducing the possibility of baseband clipping due to large motion artifacts and by improving the receiver dynamic range.

4.1.4   Nonlinear Detection of Complex Vibration Patterns

As discussed in Chapter 2, the nonlinear Doppler phase modulation effect can be used to measure both the frequency and amplitude of a sinusoidal motion. This has triggered further investigations that use the nonlinear Doppler phase modulation effect to measure more complex periodic motion that includes multiple harmonics. Intuitively, if Doppler radar can obtain the amplitude and frequency of each harmonic, the detection technique can potentially establish itself as a useful tool in vibration measurement or mechanical diagnosis (Lai, 1993). For example, when the external driving force is periodic, the resulting vibration of a single degree-of-freedom dynamic linear system, such as a simple spring-mass-damper system, will comprise frequencies determined by the frequency content of the external driving force (Rao, 1986).

To measure a vibration composed of three harmonics using nonlinear Doppler phase modulation, a wavelength division sensing RF vibrometer can be realized using a tunable carrier frequency Doppler radar sensor. In Yan et al. (2011a), by measuring the variation of the amplitude ratio of a fixed harmonic pair from the baseband spectrum for three different carrier frequencies, the amplitude of each harmonic component of the vibration can be extracted. Compared with Doppler radar that measures the frequency and amplitude of a simple sinusoidal motion, there are several factors that need to be considered. First, in order to determine the proper carrier frequency and motion amplitude to be measured, it is necessary to explore the reason why the magnitude of some harmonics in the baseband spectrum is stronger than others and how the Bessel function coefficient affects the harmonic strength. Second, a guideline in about how to choose the basic harmonic pair in measurement is necessary for practical applications. It is possible that some pairs of harmonics can achieve more accurate results than others. This section will discuss these problems using analysis and experimental results. To understand the detection mechanism, harmonic vibrations need to be modeled first.

When a subject is driven by an external force that is complex periodic, it will result in a harmonic vibration of a dynamic system. The force will lead to a steady-state output with a term of each frequency component. The free vibration without external driving force of a single degree-of-freedom system, such as a simple spring-mass-damper system, will be sinusoidal. If the external driving force is purely sinusoidal, the steady-state response of the linear system will be at the same frequency as the excitation. When the excitation force is complex periodic, such as for a square wave, the response will be a multi-tone harmonic vibration, which can be described as

$$x(t) = m_1 \sin(\omega t + \phi_1) + m_2 \sin(2\omega t + \phi_2) + \cdots$$
$$+ m_N \sin(N \omega t + \phi_N).$$

In the above equation, m_N is the amplitude of each harmonic of the vibration, ω is the fundamental frequency. As the phase angles of each motion $\phi_1, \phi_2, \ldots, \phi_N$ are determined by the initial conditions and may assume different values for every independent test, they are usually randomly distributed unless the initial conditions of the system are fixed. On the other hand, the amplitude response is determined by the characteristics of the system itself and the external driving force, such as the mass of the system (m), the spring coefficient (k), the natural frequency (ω_n) of the system, and the magnitude of the external driving force (F_0) of a spring-mass-damper vibration system.

The conventional way to obtain the amplitude and phase of each harmonic of a complex vibration in practical application is through numerical procedure, such as Simpson's rule (Gerald and Wheatley, 1984):

$$A_n = \frac{2}{M} \sum_{i=1}^{M} x_i \cos \frac{2n\pi t_i}{\tau},$$

$$B_n = \frac{2}{M} \sum_{i=1}^{M} x_i \sin \frac{2n\pi t_i}{\tau},$$

where τ is the vibration period, $t_1, t_2, \ldots, t_N$ are the equidistant sampling points with $\Delta t = \tau/N \cdot x_i$ is the corresponding value of the motion at t_i. Then the amplitude and phase can be determined as

$$\text{Amplitude: } m_n = \sqrt{A_n^2 + B_n^2},$$

$$\text{Phase: } \phi_n = \tan^{-1}\left(\frac{B_n}{A_n}\right).$$

The advantage of the numerical method is that it can obtain both the amplitude and phase of each harmonic, which means it can recover the pattern of the vibration. However, this can be guaranteed only when there are sufficient time-domain samples. Otherwise, the recovered pattern will also be distorted. On the other hand, instead of using the time-domain data, the radar sensor using nonlinear phase modulation utilizes the frequency-domain information. In particular, the amplitude ratio between harmonics in the baseband spectrum is used to extract the amplitude and frequency of each frequency component of a vibration, resulting in the concept of "RF vibrometer". RF vibrometer can be a helpful tool in mechanical testing and diagnose. For example, the response spectrum of a motor under an external square-wave driving force that contains multiple harmonics of a fundamental frequency reveals the natural frequency, because the peak of the spectrum represents where resonance happens. From the spectrum under a sinusoidal driving force, people can also check if a lathe in an assembly line is in good condition structurally and the resistance against external vibration during delivery. The frequency of interest is around $0-100$ Hz.

Two techniques can be used to detect complex motion pattern based on nonlinear phase modulation. They can be understood using the following example that detects a vibration with two harmonics

$$x(t) = 3\sin(2\pi f_1 t) + 2\sin(2\pi f_2 t).$$

Using complex signal demodulation introduced in Chapter 2, the complex baseband I/Q signals can be obtained as

$$S(t) = I(t) + iQ(t) = \exp\left(i\left[\frac{4\pi x(t)}{\lambda} + \phi\right]\right)$$

$$= \sum_{p=-\infty}^{\infty} \sum_{l=-\infty}^{\infty} J_p\left(\frac{4\pi m_1}{\lambda}\right) J_1\left(\frac{4\pi m_2}{\lambda}\right) e^{i[2\pi(f_1 p + f_2 l)t]} \cdot e^{i\phi},$$

where ϕ is the total residual phase and λ is the carrier wavelength. $J_n(x)$ represents the first kind Bessel function of the nth order. As $e^{j\phi}$ has a constant-envelope of unity, the effect of ϕ on signal amplitude is thus eliminated, leaving the amplitude of the harmonic determined only by the Bessel function coefficient. When the term $f_1 p + f_2 l = x$, it denotes the harmonic frequency equaling to x Hertz, and its strength will be

$$H_x = \left| \sum_{p=-\infty}^{\infty} \sum_{l=(x-f_1 p)/f_2}^{\infty} J_p\left(\frac{4\pi m_1}{\lambda}\right) J_l\left(\frac{4\pi m_2}{\lambda}\right) \right|.$$

It can be seen that the amplitude ratio between two harmonics is a function of λ, which can be represented as $H_x/H_y = f(\lambda)$.

4.1.4.1 Wavelength Division Sensing
The wavelength division sensing technique detects the complex motion with multiple carrier frequencies. It works in the following steps.

Step 1: Choosing two harmonics H_x and H_y
Step2: Measuring H_x/H_y at carrier frequency f_{c1} to obtain
$H_x/H_y = f_1(\lambda_1)$
Step3: Measuring H_x/H_y at carrier frequency f_{c2} to obtain
$H_x/H_y = f_2(\lambda_2)$
Step 4: m_1 and m_2 can be obtained by solving two equations.

4.1.4.2 Detection Using Multiple Harmonic Pairs with a Single Carrier Frequency
The detection accuracy of the wavelength division sensing technique strongly depends on the selection of harmonic pairs. Without inspecting the amplitude expression of harmonic on the baseband spectrum, there may be large deviation of the final calculated amplitude of each harmonic motion consisting of the vibration. In the meantime, as the carrier frequency needs to be tunable, it requires a frequency synthesizer or wide tuning range VCO and a broadband antenna, all of which will increase the complexity of circuit design

and the manufacturing cost as well. Detection using multiple harmonic pairs solves the problem with a fixed carrier frequency. It works in the following steps.

Step 1: Choosing two different harmonic pairs (H_x and H_y) and (H_y and H_z)

Step 2: Measuring the harmonic amplitude ratios at a fixed carrier frequency f_c

$$H_x/H_y = f_1(\lambda_1) \quad \text{and} \quad H_u/H_v = f_2(\lambda_1)$$

Step 3: m_1, m_2 can be obtained by solving two equations

This technique can be easily extended to detect vibration with more than two harmonics. As this method uses multiple harmonic pairs, the probability of all harmonic pairs being inappropriate is small.

4.1.4.3 *Experimental Demonstration* A 5.8 GHz RF vibrometer based on the detection using multiple harmonic pairs is implemented on PCB level. Two patch antennas with 9 dB gain are connected to the vibrometer board through a connector. The moving target is driven by a linear actuator that can be controlled by a laptop program. The experimental setup of the vibration detection is shown in Fig. 4.13. The thickness of the wall is 15 cm and detection distance is 1.5 m. The parameters of amplitude, frequency, and phase angle of each tone of the vibration can be set in the program. The sampling rate is 20 Hz, and the fast Fourier transform (FFT) window size is 512.

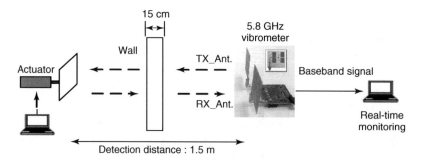

Figure 4.13 Experiment setup of the see-through-wall vibration detection.

Besides displaying the real-time radar baseband output and spectrum, Matlab codes are integrated into a LabView program to realize real-time movement pattern monitoring. A peak search block is used in LabView to collect the peaks on the spectrum and their corresponding frequencies. The data is then transferred into another block that runs the detection algorithm. When the background calculation is finished, the vibration pattern of the target will be shown on the user interface immediately.

Figure 4.14 shows the real-time monitored pattern of two-tone and three-tone vibrations recovered from multiple harmonic pairs with a single carrier frequency. The excess phase angles are set to be 0 in

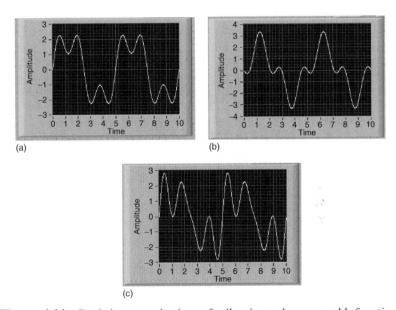

(a)   (b)

(c)

Figure 4.14 Real-time monitoring of vibrations that are odd functions in time using RF vibrometer. (a) Two-tone movement with programmed values of $m_1 = 2$ mm, $m_2 = 1$ mm, $f_1 = 0.2$ Hz, and $f_2 = 0.6$ Hz; measurement result: $m_1 = 2.1$ mm, $m_2 = 1.05$ mm, $f_1 = 0.2$ Hz, and $f_2 = 0.6$ Hz. (b) Two-tone movement with programmed values of: $m_1 = 2$ mm, $m_2 = -1$ mm, $f_1 = 0.2$ Hz, and $f_2 = 0.6$ Hz; measurement result: $m_1 = 2.15$ mm, $m_2 = -1.1$ mm, $f_1 = 0.2$ Hz, and $f_2 = 0.6$ Hz. (c) Three-tone movement with programmed values of: $m_1 = 1.8$ mm, $m_2 = 1.2$ mm, $m_3 = 0.8$ mm, $f_1 = 0.2$ Hz, $f_2 = 0.6$ Hz, and $f_3 = 0.8$ Hz; measurement result: $m_1 = 1.77$ mm, $m_2 = 1.2$ mm, $m_3 = 0.84$ mm, $f_1 = 0.2$ Hz, $f_2 = 0.6$ Hz, and $f_3 = 0.8$ Hz.

these cases. The measurement values and the programmed values of each case are reported in the caption. The deviations are all within 10%. Thus, the improved monitoring program has fulfilled in real time the function of providing the movement pattern of an unknown vibration that is odd function in time (zero phase angles). It is worth mentioning that for a typical commercial quadrature mixer, it has been verified that even if it works at the worst condition that has both the maximum amplitude mismatch and maximum phase mismatch, the resulting detection error would be less than 10% (Yan et al., 2010). For quadrature mixer with normal functionality, the introduced error from I/Q imbalance should be negligible (Yan et al., 2010).

The sensitivity of the harmonic ratios to the phase angle has also been verified in the experiment. Table 4.1 shows the real time measured m_1 and m_2 using H_1/H_3 and H_3/H_4 at different combinations of the additional phase angle ϕ_1 and ϕ_2 set in the program. Table 4.2 lists the measurement results by using H_2/H_3 and H_3/H_4. Compared to the programmed value of $m_1 = 2$ mm and $m_2 = 1$ mm, it can be seen that by using insensitive harmonic ratio pairs, results with much less deviation can be obtained.

More examples and discussions of the complex movement pattern monitoring technique are reported in Yan et al. (2011b) The harmonic analysis based on the combination of Bessel function coefficients

TABLE 4.1   Measured m_1 and m_2 Using H_1/H_3 and H_3/H_4

		$\pi/6$	$\pi/4$	$\pi/2$	$5\pi/3$	2π
	m_1			ϕ_1		
ϕ_2	$\pi/6$	2.0783	2.1752	2.171	2.2014	2.2862
	$\pi/4$	2.2911	2.2869	2.0757	2.0033	1.8986
	$\pi/2$	1.8963	2.3001	2.3053	2.2736	2.0876
	$5\pi/3$	2.3017	1.8988	2.087	2.1958	2.2142
	2π	1.8723	2.2749	1.9866	1.8782	1.7701
	m_2			ϕ_1		
		$\pi/6$	$\pi/4$	$\pi/2$	$5\pi/3$	2π
ϕ_2	$\pi/6$	1.0548	1.1344	1.05	1.2386	1.086
	$\pi/4$	0.8369	0.9363	0.8347	0.769	1.0381
	$\pi/2$	0.7881	0.8188	1.1396	1.0746	1.2367
	$5\pi/3$	0.8387	0.7685	1.2366	0.8379	1.0651
	2π	1.1338	1.2443	0.8364	1.0749	0.8334

TABLE 4.2 Measured m_1 and m_2 Using H_2/H_3 and H_3/H_4

		$\pi/6$	$\pi/4$	$\pi/2$	$5\pi/3$	2π
	m_1			ϕ_1		
ϕ_2	$\pi/6$	3.6271	3.382	2.2689	4.4606	4.0168
	$\pi/4$	4.1694	4.0179	3.3829	4.503	4.3909
	$\pi/2$	4.295	4.3926	4.5193	3.0913	4.0181
	$5\pi/3$	4.461	4.3909	4.0172	4.2948	4.5177
	2π	3.0905	3.3806	4.0168	3.6262	2.2671
	m_2			ϕ_1		
		$\pi/6$	$\pi/4$	$\pi/2$	$5\pi/3$	2π
ϕ_2	$\pi/6$	0.5811	0.6558	1.135	−1.2165	0.4843
	$\pi/4$	0.4531	0.4844	0.6559	−1.151	−1.3415
	$\pi/2$	−1.5663	−1.3441	−1.1311	0.7604	0.4848
	$5\pi/3$	−1.2167	−1.343	0.4847	−1.5637	−1.1301
	2π	0.7595	0.6561	0.4846	0.5814	1.1349

suggests that measuring multiple harmonic pairs at a fixed carrier frequency could be a reliable approach for complex motion pattern measurement. Compared with the wavelength division sensing technique, it reduces the probability of ending up with results that have big errors caused by the wrong choice of a single inappropriate harmonic pair. The analysis also establishes itself as a guideline on how to improve the detection accuracy and reliability by first inspecting the Bessel function coefficient of each harmonic. When the vibration contains both odd function and even function harmonics (phase angles are not zero), by picking the insensitive harmonics from the baseband spectrum, the amplitude of each frequency component can still be recovered, from which the characteristics of the system can be identified.

4.1.5 Motion Sensing Based on Self-Injection-Locked Oscillators

Injection locking is an interesting nonlinear phenomenon in physics that has been studied for a very long period of time since the seventeenth century. Even in recent years, its applications in modern electronic circuits are still being explored (Razavi, 2004). The first report of such effect in electronic oscillators was written in German by Möller in 1921 (Moller, 1921; Mackey, 1962). When a sinusoidal signal is injected into an oscillator, the oscillation frequency of the oscillator will be locked to the frequency of the injection signal if these two frequencies are close enough. The theory and models to explain the injection locking in oscillators have been developed by many researchers including Adler (1946), Huntoon and Weiss (1947), Paciorek (1965), and Kurokawa (1973).

In addition to many applications in frequency synthesizers, the injection locking can also be used for spectrum sensing and cognitive radio applications (Li et al., 2009d). Figure 4.15 illustrates the block diagram of a spectrum sensing system using injection locking. Signals in the air are received by antenna, amplified by the LNA, and injected into the VCO. By sweeping the oscillation frequency of the VCO through its tuning range, which covers the portion of spectrum to be detected, the oscillator will go through injection pulling and injection locking when its oscillation frequency gets close to the injected signal. After frequency demodulation and bandpass filtering, a "quiet zone" reveals the frequency of detected signal.

Instead of detecting wireless signals in the air by using the injection locking method as described above, if the oscillator transmits its signal through an antenna to hit a moving target, and the reflected signal is used as the injection signal to the oscillator, the system may detect the motion of the target modulating on the reflected wave due to Doppler effect. Vital signs of a human subject can therefore be detected by this *self-injection-locking* technique. Figure 4.16 illustrates this concept (Wang et al., 2010). Combining the circuits and algorithms used for spectrum sensing, a system capable of concurrent spectrum sensing and vital sign sensing can be realized. Figure 4.17 shows the detailed block diagram of such a system. It is modified from the system in

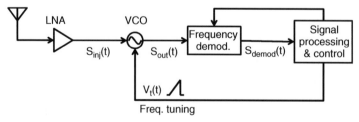

Figure 4.15 Block diagram of RF spectrum sensing system. Adapted from Li et al., (2009d).

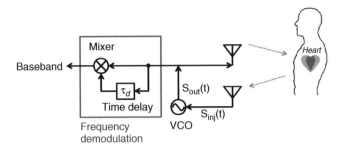

Figure 4.16 Self-injection-locking technique for noncontact vital sign detection. Adapted from Wang et al., (2010).

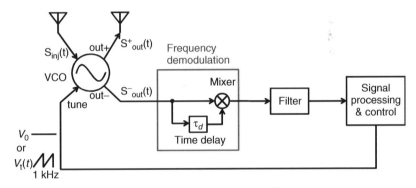

Figure 4.17 Block diagram of the concurrent spectrum sensing and non-contact vital sign sensing. Adapted from Wang et al., (2010).

Fig. 4.15 by using a differential oscillator and taking one of its differential outputs to transmit it through the transmitting (TX) antenna. The baseband output circuit is also modified to perform both vital sign sensing and spectrum sensing. The motivation of combining both functions is to detect possible wireless interference and select a radar carrier frequency avoiding interference. It should be noted that in this self-injection-locking technique, the null detection points exist when the phase delay between the injected signal and the oscillator output signal is an odd multiple of 90°. However, the frequency sweeping alleviates this problem. Figure 4.18 shows the measured heartbeat signals at a null detection point with and without frequency sweeping turned on. A simultaneously measured ECG data is also plotted for comparison. If the operating frequency is fixed, a phase shifter in the

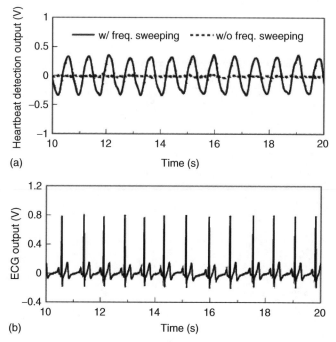

(a)

(b)

Figure 4.18 (a) Noncontact detection of heartbeat using self-injection-locking technique. The signal was measured at a null detection point. With frequency sweeping, heartbeat is clearly detected. (b) Simultaneously measured ECG output to verify the detection accuracy. Copyright © IEEE. All rights reserved. Reprinted with permission from Wang F-K, Li C-J, Hsiao C-H, Horng T-S, Lin J, Peng K-C, Jau J-K, Li J-Y, Chen C-C. A novel vital-sign sensor based on a self-injection-locked oscillator. IEEE Trans Micro Theory Tech 2010; 58(12):4112–4120.

receiver path can be used to tune the phase of injected signal out of null detection points. Operating at 1.8 GHz with about 1 mW output power, the maximum sensing distance is 1 m. According to the analysis in Wang et al. (2010), doubling the operating frequency will quadruple the maximum sensing distance of this self-injection-locking architecture. The maximum sensing distance increases to 4 m when the operating frequency is increased to 3.6 GHz.

The self-injection-locking motion sensing radar can use a single antenna for both transmitting and receiving instead of two antennas for separate transmission and reception. This is achieved by using one of the oscillator's differential output ports connected to antenna as the

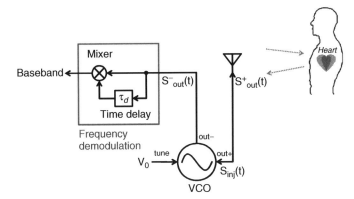

Figure 4.19 Single-antenna self-injection-locking motion sensing radar. Adapted from Wang et al., (2011).

injection port as well. The structure is shown in Fig. 4.19 (Wang et al., 2011). The clutter signal due to reflection from the antenna causes an additional phase delay in the received Doppler signal $S_d(t)$ and subsequently changes the SNR gain resulted from injection locking process, as shown in Fig. 4.20. It can be seen from the figure that the clutter does not affect the maximum SNR gain, and it effectively

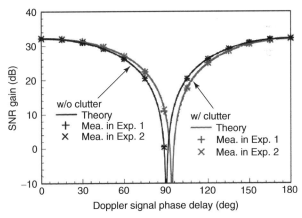

Figure 4.20 SNR gain versus Doppler signal phase delay ϕ_d. Copyright © IEEE. All rights reserved. Reprinted with permission from Wang F-H, Horng T-S, Lin J, Peng K-C, Jau J-K, Li J-Y, Chen C-C. Single-antenna Doppler radars using self and mutual injection locking for vital sign detection with random body movement cancellation. IEEE Trans Micro Theory Tech 2011; 59(12):3577–3586.

shifts the null detection point. Therefore, by adjusting the operating frequency or the distance between radar and the subject, the maximum SNR gain still can be achieved.

Similar to the random body movement cancellation using two radars described in Section 4.1.3, two self-injection-locking motion sensing radars can be placed at opposite sides of a human subject to cancel the random body movement artifacts and extract the vital signs accurately. In this system, two self-injection-locking radars are operating at similar frequencies. The signal from one of the radars will reach the other radar and the two radars are eventually locked to each other through mutual injection locking. The random body movement is cancelled when the difference between the round-trip phase delays of the two radars equals to any multiple of 2π, which can be realized by adjusting the radar operating frequency or the subject's position between the two radars. The baseband signal can be collected from either radar. Using this technique, measurement of a subject's vital signs while jogging on a treadmill was successfully demonstrated, for the first time, in 2011 (Wang et al., 2011).

4.2   NUMERICAL METHODS: RAY-TRACING MODEL

In real applications of radar motion detection, a real antenna with a certain radiation pattern does not have infinite directivity. Therefore, signals are reflected off and captured from different parts of the body. When signals on different paths with different intensity and residual phases are received by the radar, they are simply summed together by the receiving antenna, either canceling out or enhancing the desired signal components. Therefore, a ray-tracing model was developed to compensate for the shortage of the single-beam model that we have been using so far.

As shown in Fig. 4.21, the actual received signal should be represented from a ray-tracing point of view as

$$
I(t) = \iint_{s} E(x, y) \cdot \cos\left[\Delta\phi + \frac{4\pi}{\lambda} \cdot \left\{ \rho(x, y)^2 \right.\right.
$$
$$
\left.\left. + \left[d_0 + m_h(x, y)\sin(\omega_h t) + m_r(x, y)\sin(\omega_r t) \right]^2 \right\}^{1/2} \right] ds,
$$

$$(4.10)$$

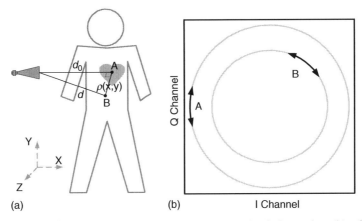

Figure 4.21 Ray-tracing model (a) and the angular information (b) of signals reflected from point A and B on the body using a 5.8 GHz radar. The antenna is facing the body in the $-Z$ direction of the X–Y–Z coordinate (Li et al., 2008a).

$$Q(t) = \iint\limits_{s} E(x, y) \cdot \sin\left[\Delta\phi + \frac{4\pi}{\lambda} \cdot \left\{\rho(x, y)^2\right.\right.$$

$$\left.\left. + \left[d_0 + m_h(x, y)\sin(\omega_h t) + m_r(x, y)\sin(\omega_r t)\right]^2\right\}^{1/2}\right] ds,$$

$$(4.11)$$

where $E(x, y)$ is the intensity of the electromagnetic wave reflected from the location (x, y) of the human body, $\Delta\phi$ is the residual phase shift accumulated in the electronic circuit, $\rho(x, y)$ is the distance between location (x, y) and the antenna's projection on the chest wall, and d_0 is the distance between the antenna and the subject under test. The integration is carried out over the entire surface of the subject under test.

Assume the antenna is placed 1 m away in front of the heart center, and the locations of the heart center A and the body center B on the front chest wall are separated by 11 cm. The difference in the transmission path for signals from the antenna to the two points is $\Delta x = \sqrt{1^2 + 0.11^2} - 1 = 0.006$ m, which is replicated in the receiving path and would produce a phase difference of $83.5°$ for a 5.8 GHz radar. Meanwhile, the radiation intensity of the antenna on the body surface is different from point to point, depending on the antenna radiation pattern. This implies that the received baseband signals from the two points will have two different loci in the constellation graph, as shown in Fig. 4.21b. Therefore, the actual vital sign detection is complicated by the phase offset and the accuracy might be affected. Numerical simulations are needed for further analysis. An example is given as follows.

Assuming the antenna is 1 m in front of the heart center and the subject is 1.8 m high, the phase offset in different paths compared with the beam propagating to the center of the heart is shown in Fig. 4.22a for a 5.8 GHz radar sensor. Dramatic change in phase offset is observed. Figure 4.22b represents the radiation intensity on the human body produced by an ideal 7 × 7 antenna array comprising omnidirectional antennas spaced by λ/2. Figures 4.22c and d show the approximation of the normalized amplitude of body movements caused by respiration and heartbeat, respectively. It can be inferred that when a carrier frequency of 24 GHz is used for the higher sensitivity at shorter wavelengths, more significant phase change will be observed.

To demonstrate the difference between a simplified single-beam model and the ray-tracing model, the optimal carrier frequency for noncontact vital sign detection will be investigated again here. Authors are encouraged to compare the results in this section with the results of Chapter 2, where a single-beam model was established in ADS for the same problem.

Assuming detecting from the front of a human body with respiration-induced body movement amplitude (peak value) of $m_h = 1.2$ mm and heartbeat-caused body movement amplitude (peak value) of $m_r = 0.15$ mm, the received baseband signal can be generated based on the ray-tracing equations given in this section. Analyzing the spectrum of simulated baseband signal for a person seated 1 m in front of the sensor, the detected heartbeat signal strength as a function of RF is shown in Fig. 4.23, where different curves correspond to different antenna beamwidth. Two-dimensional uniform antenna arrays with different numbers of omnidirectional antennas were used to obtain desired beamwidth. And the total input power to the antenna array was kept the same. The relative strength of heartbeat signal compared with respiration signal is shown in Fig. 4.24.

It is shown in Fig. 4.23 that the absolute strength of heartbeat signal increases with RF until it reaches a relatively flat region, after

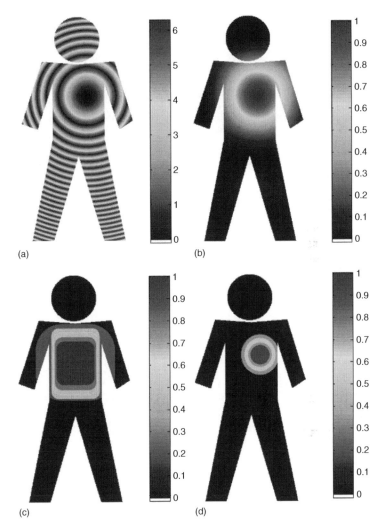

Figure 4.22 Ray-tracing model. (a) The phase offset on the surface of human body radiated by a 5.8 GHz radar; (b) A 7×7 elements antenna array's radiation intensity on the human body. (c) Approximation of the normalized amplitude of body movement caused by respiration. (d) Approximation of the normalized amplitude of body movement caused by heartbeat. Copyright © IEEE. All rights reserved. Reprinted with permission from Li C, Ling J, Li J, Lin J. Accurate Doppler Radar Non-contact Vital Sign Detection Using the RELAX Algorithm. IEEE Transactions on Instrumentation and Measurement 2010; 59(3):687–695.

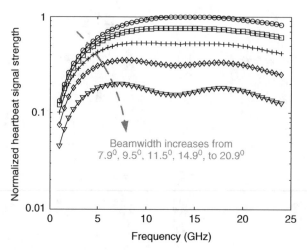

Figure 4.23 Detected heartbeat signal strength as a function of radio frequency for different antenna beamwidth (Li et al., 2007c).

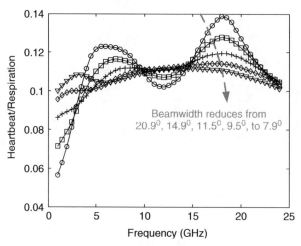

Figure 4.24 Relative heartbeat signal strength compared with respiration signal, as a function of radio frequency for different antenna beamwidth (Li et al., 2007c).

which the signal strength begins to decrease. Meantime, high directivity antenna is desirable for two reasons: First, low antenna directivity causes a near field effect. Signals with different phases due to large beamwidth will cancel each other, making the desired signal strength fluctuate with RF, as shown by the curves with beamwidth of $20.93°$ and $14.85°$ in Fig. 4.23 and Fig. 4.24. This corresponds to a detected signal level fluctuation for people with different body movement amplitude. Second, large antenna directivity effectively enhances detected signal strength. Therefore, simulation based on the ray-tracing model demonstrates that a RF from 5 GHz to the lower region of Ka-band is reasonable for vital sign detection and an antenna with appropriate directivity could be used for optimal detection.

4.3 SIGNAL PROCESSING

Signal processing plays a very important role in radar. Once the signal in baseband is digitized by an ADC, signal processing is needed to filter undesired noise and interference and extract wanted information from vital sign or vibration detection. This may involve extracting multiple periodic signals with different frequencies, which may be time-variant and containing harmonics as well. This section introduces a few techniques useful for the above purposes.

4.3.1 MIMO and SIMO Techniques

Besides the undesired motion of the radar and the subject to be detected, the other major problems with radar noncontact motion sensing are the noise caused by motion artifacts and the presence of multiple subjects. The MIMO architecture was then applied to solve these problems. In Boric-Lubecke et al. (2005), techniques were proposed based on the Bell-Labs LAyered Space-Time (BLAST) solution to isolate useful scattered radio signals from those due to unwanted motion in the environment. First, if there is interfering motion in the radar field of view, multiple antennas can be used to detect multiple copies of the same signal with different phase information. Coherent combining will then be used to provide a greatly improved estimate of desired motion characteristics. It was also proposed that when more than one target is in view, multiple transmitters and receivers providing multiple copies of signal could be used to distinguish between the different sources of Doppler motion, isolate the desired signal, and determine a number of targets.

Simulations were performed for a MIMO system with four transmitting and four receiving antennas. The antennas have a carrier frequency of 2.4 GHz, 2λ separation between transmit elements, and $\lambda/2$ separation between receive elements. Simulation results have demonstrated the effectiveness of the MIMO technique for enhancing detection accuracy when clutter and multiple targets are present. However, because of the complexity of experimental setup and significant computational load, not many continued reports were found on real-world implementation.

Another proposed approach is the single-input, multiple-output (SIMO) technique. In Zhou et al. (2006) and Boric-Lubecke et al. (2005), the single and multiple antenna systems and SIMO/MIMO signal processing were explored to isolate desired radar return signals from multiple subjects. It has been experimentally demonstrated that up to two subjects can be separated in a single-antenna system. Simulations have also shown that in case two subjects have identical cardiovascular behavior, it is still possible to distinguish those signals using MIMO techniques. Also, a generalized likelihood ratio test (GLRT), based on a model of the heartbeat, has been developed to show that this technique can be used to distinguish between the presence of two, one, or zero subjects, even with a single antenna. Furthermore, this technique was extended to detect up to $2N - 1$ subjects using N antennas. Although, SIMO/MIMO is a promising solution to motion artifacts and the problem happened when multiple subjects exist, limited experimental results have been reported so far. In the meantime, if there is heavy multipath effect in the system, some of the current methods cannot be used. Instead, blind source separation methods have to be used for motion sensing in heavy multipath environment (Petrochilos et al., 2007).

4.3.2   Spectral Estimation Algorithms

As discussed in Chapter 2, harmonics and intermodulation interferences generated from the nonlinear Doppler phase modulation effect complicate the radar baseband output spectrum and degrade the detection accuracy. In the meantime, all the frequency components are surrounded by clutter noise on the spectrum. Therefore, the task of estimating sinusoidal frequencies from the received measurements has to be accomplished by a reliable and efficient spectral estimation algorithm.

The most widely used frequency estimation approaches for radar noncontact motion sensing are based on the periodogram obtained from FFT. Once the nonparametric periodogram is derived from the Fourier transform, the estimated frequency tones correspond to the locations of the dominant peaks of the periodogram. However, the periodogram suffers from smearing and leakage problems. Smearing decreases the resolution, making two closely spaced sinusoids irresolvable. Take noncontact vital sign detection as an example, the third- or fourth-order harmonics of respiration might be located near the heartbeat signal on the spectrum. Consequently, their mainlobes merge into a single peak because of smearing. In addition, the sidelobe of a strong signal can bury the weak signals because of leakage. Typically, the respiration signal dominates the spectra, and its harmonics may overwhelm the heartbeat signal, making the latter invisible in the periodogram. In either situation, the periodogram based on Fourier transform may not provide sufficient detection accuracy.

An alternative to periodogram is the parametric maximum-likelihood (ML) estimator, which can be much more accurate than the periodogram. However, with a large number of sinusoids (a situation encountered in the motion sensing application), the ML estimator becomes computationally prohibitive because of the need to search a large dimensional space.

Recently, a parametric and cyclic optimization algorithm, referred to as the *relaxation (RELAX) algorithm*, has demonstrated its value for the baseband signal processing in radar motion sensing. The RELAX algorithm, originally proposed by Li and Stoica (1996), is an asymptotic ML approach and significantly outperforms the conventional periodogram. RELAX iteratively estimates the parameters of each sinusoid in a "super clean" fashion. As long as the stronger sinusoids are accurately estimated, RELAX is able to reveal the weak sinusoids by subtracting them out from the measurements. The rest of this section will use vital sign detection as an example to demonstrate the advantage of RELAX as compared with ordinary periodogram.

On the basis of the analysis in Section 2.3.5, the baseband output complex signal can be represented as

$$B(t) = \sum_{k=-\infty}^{\infty} \sum_{l=-\infty}^{\infty} J_l\left(\frac{4\pi m_h}{\lambda}\right) J_k\left(\frac{4\pi m_r}{\lambda}\right) e^{j(k\omega_r t + l\omega_h t + \phi)} + v(t)$$

$$= \sum_{k=-\infty}^{\infty} \sum_{l=-\infty}^{\infty} C_{k,l} e^{j\phi} \cdot e^{jt\omega_{k,l}} + v(t), \tag{4.12}$$

where $v(t)$ represents the noise term. As $e^{j\Phi}$ has a constant unity modulus, the null detection point problem due to the effect of Φ on the signal amplitude is eliminated. Therefore, $e^{j\Phi}$ will be ignored in the following analysis.

In the following, the detection will be formulated as a spectral estimation problem. Vectors and matrices will be denoted by bold-face lowercase and uppercase letters. $(\ )^T$ and $(\ )^H$ will represent transpose and conjugate transpose, respectively. $\|\ \|$ will represent the Euclidean norm in this section. And the ith component of a vector $\mathbf{b}$ will be written as b_i. A sampled version of the baseband output can be written as

$$B(n\Delta_t) = \sum_{k=-\infty}^{\infty} \sum_{l=-\infty}^{\infty} \beta_{k,l} \cdot e^{jn\Delta_t \omega_{k,l}} + v(n\Delta_t), \tag{4.13}$$

where $n = 0, 1, \ldots, N-1$ with N being the sampled data length, and Δ_t is the sampling interval. The complex-valued amplitude $\beta_{k,l}$ is equal to $C_{k,l} e^{j\Phi}$. Denoting the normalized frequency as $\widetilde{\omega}_{k,l} = \omega_{k,l}$, the baseband output can be simplified as

$$B_n = \sum_{k=-\infty}^{\infty} \sum_{l=-\infty}^{\infty} \beta_{k,l} \cdot e^{jn\widetilde{\omega}_{k,l}} + v_n. \tag{4.14}$$

The sequence $\{B_n\}_{n=0}^{N-1}$ in the Equation 4.14 consists of an infinite number of sinusoidal components, and it is infeasible to estimate an infinite number of unknown parameters (i.e., $\beta_{k,l}$'s and $\widetilde{\omega}_{k,l}$'s) from a finite number of data samples. Fortunately, the power of the harmonics decays very quickly, such that it can be assumed that $\{B_n\}_{n=0}^{N-1}$ is composed of K strongest sinusoids by neglecting the weak sinusoids. The value of K will be chosen so that the desired sinusoidal components are not omitted. Mapping $\beta_{k,l}$'s and $\widetilde{\omega}_{k,l}$'s of the K strongest sinusoids to $\{\alpha_k\}_{k=0}^K$ and $\{\theta_k\}_{k=0}^K$ respectively, Equation 4.14 can be simplified as

$$B_n = \sum_{k=1}^K \alpha_k e^{j\theta_k n} + \tilde{v}_n, \tag{4.15}$$

where $\tilde{v}_n$ is the sum of the noise term v_n and the neglected weak sinusoidal components.

Denoting the received symbol vector as $y = [y_0 \; y_1 \cdots y_{N-1}]^T$, the steering vector as $\mathbf{a}(\theta_k) = [e^{j\theta_k} \cdots e^{j(N-1)\theta_k}]^T$, and the noise vector as $\mathbf{v} = [\tilde{v}_0 \; \tilde{v}_1 \; \cdots \; \tilde{v}_{N-1}]^T$, Equation 4.15 can be further expressed as

$$\mathbf{B} = \sum_{k=1}^K \alpha_k \mathbf{a}(\theta_k) + \mathbf{v}. \tag{4.16}$$

Therefore, the problem is reduced to estimating $\{\alpha_k, \theta_k\}_{k=1}^K$ given $\mathbf{B}$. Two approaches (i.e., periodogram and RELAX) will be compared in the following.

4.3.2.1 Periodogram. The periodogram is defined as

$$P(\theta) = \frac{1}{N} \left| \sum_{n=0}^{N-1} B_n e^{-j\theta n} \right|^2 = \frac{\|\mathbf{a}^H(\theta)\mathbf{B}\|^2}{N}. \tag{4.17}$$

The FFT speeds up the computation of the periodogram, and zero-padding y before conducting FFT is essential to achieve high frequency estimation accuracy.

Ideally, with an infinite sample length (i.e., $N \rightarrow \infty$), the kth sinusoidal component contributes one vertical line in the spectrum at frequency θ_k. Practically, however, the finite data length N causes smearing and leakage in the periodogram. Figure 4.25 shows an example where the time-domain data comprises four complex-valued sinusoids without noise. The data length is $N = 40$, and the sampling frequency is 1 Hz. Stems represent the true line spectra, and the curve represents the periodogram of $\{B_n\}_{n=0}^{39}$. The small number of N smears the line spectrum and merges two closely spaced sinusoids into one peak. More specifically, the sinusoids at -0.115 and -0.1 Hz

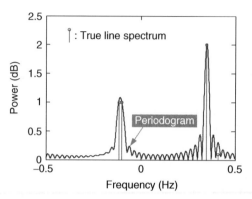

Figure 4.25 Four complex-valued sinusoids with frequencies of -0.115, -0.1, 0.35, and 0.412 Hz are presented with data length $N = 40$ and a sampling frequency of 1 Hz (Li et al., 2010a).

are located within the resolution limit of the periodogram (i.e., $1/N$ Hertz). Consequently, their mainlobes superpose to form a single peak.

In the meantime, sidelobes are generated because of leakage. In this example, the sidelobe of the strong signal (at 0.35 Hz) buries the weak signal at 0.412 Hz, making it almost invisible in the periodogram. If $K = 4$ is known as the *a priori* information, then from the periodogram shown in Fig. 4.25, it is tempting to claim the frequency estimates as 0.35, -0.107, 0.314, and 0.386 Hz (the last two values correspond to the sidelobe peaks of the 0.35 Hz signal). Obviously, these estimates are unacceptable.

For realistic vital sign detection, more challenges are present. For example, the signals are measured in noisy environments. When the noncontact Doppler radar is used for real-time portable applications, such as earthquake or fire emergency rescue, the sample length N cannot be too long (typically in hundreds) as a large N imposes considerable memory and computational loads. In addition, there exists a significant probability that the harmonics of respiration signal are close to the heartbeat signal. Furthermore, the respiration signal is often much stronger than the other sinusoidal components. To address these challenges, a more reliable spectral estimation algorithm is desired.

4.3.2.2 *RELAX algorithm.*   In the presence of zero-mean white Gaussian noise, the ML estimator requires solving the following non-linear least-squares fitting problem

$$
\{\widehat{\alpha}_k, \widehat{\theta}_k\}_{k=1}^K = \arg \min_{\{\alpha_k, \theta_k\}_{k=1}^K} \left| B_n - \sum_{k=1}^K \alpha_k e^{j\theta_k n} \right|^2. \tag{4.18}
$$

The above problem can be solved efficiently by employing the RELAX algorithm. Its pseudocode is outlined in Table 4.3. The RELAX algorithm provides excellent initial conditions for the next new sinusoidal parameter estimation, as the estimate of a new sinusoidal component is postponed until the already determined components are "good enough."

TABLE 4.3   The RELAX Algorithm

Initialize

$$\widehat{\alpha}_k = 0, \quad k = 1, 2, \ldots, K$$

for $i = 1, 2, \ldots, K$

$$\widehat{\mathbf{B}}_i = \mathbf{B} - \sum_{k=1, k\neq i}^K \widehat{\alpha}_k \mathbf{a}(\widehat{\theta}_k)$$

$$\widehat{\theta}_i = \arg \max_\theta \frac{\|\mathbf{a}^H(\theta)\widehat{\mathbf{B}}_i\|^2}{\|\mathbf{a}(\theta)\|^2}$$

$$\widehat{\alpha}_i = \frac{\mathbf{a}^H(\widehat{\theta}_i)\widehat{\mathbf{B}}_i}{\|\mathbf{a}(\widehat{\theta}_i)\|^2}$$

repeat

for $j = 1, 2, \ldots, i$

$$\widehat{\mathbf{B}}_j = \mathbf{B} - \sum_{m=1, m\neq j}^i \widehat{\alpha}_m \mathbf{a}(\widehat{\theta}_m)$$

$$\widehat{\theta}_j = \arg \max_\theta \frac{\|\mathbf{a}^H(\theta)\widehat{\mathbf{B}}_j\|^2}{\|\mathbf{a}(\theta)\|^2}$$

$$\widehat{\alpha}_j = \frac{\mathbf{a}^H(\widehat{\theta}_j)\widehat{\mathbf{B}}_j}{\|\mathbf{a}(\widehat{\theta}_j)\|^2}$$

end for

until (convergence)

end for

Table 4.3 shows that the building block of the RELAX algorithm is to estimate one sinusoidal component with the parameter $\{\alpha_i, \theta_i\}$ from measurement vector $\widehat{\mathbf{B}}_i$ by assuming that $\widehat{\mathbf{B}}_i$ consists of this sinusoid and the noise term only. The $\{\alpha_i, \theta_i\}$ can be estimated as

$$\{\widehat{\alpha}_i, \widehat{\theta}_i\}_{k=1}^{K} = \arg \min_{\{\alpha_i, \theta_i\}} \|\widehat{\mathbf{B}}_i - \alpha_i \mathbf{a}(\theta_i)\|^2. \tag{4.19}$$

Minimizing the right side of the Equation 4.19 with respect to α_i yields

$$\widehat{\alpha}_i = \frac{\mathbf{a}^H(\theta_i)\widehat{\mathbf{B}}_i}{\|\mathbf{a}(\theta_i)\|^2}. \tag{4.20}$$

Replacing the α_i leads to

$$\widehat{\theta}_i = \arg \min_{\theta_i} \|\mathbf{P}^{\perp}_{\mathbf{a}(\theta_i)}\widehat{\mathbf{B}}_i\|^2$$

$$= \arg \max_{\theta_i} \frac{\|\mathbf{a}^H(\theta_i)\widehat{\mathbf{B}}_i\|^2}{\|\mathbf{a}(\theta_i)\|^2} \tag{4.21}$$

where $\mathbf{P}^{\perp}_{\mathbf{a}(\theta_i)} = \mathbf{I} - (\mathbf{a}(\theta_i)\mathbf{a}^H(\theta_i)/\|\mathbf{a}(\theta_i)\|^2)$ denotes the orthogonal projection onto the null space of $\mathbf{a}^H(\theta_i)$. Recalling $\|\mathbf{a}(\theta_i)\|^2 = N$ and the definition of periodogram given before, the cost function involved in the last expression of Equation 4.21 is exactly the periodogram of $\widehat{\mathbf{B}}_i$. Consequently, $\widehat{\theta}_i$ can readily be obtained as the location of the dominant peak in the periodogram.

By checking the difference of the cost function between two successive iterations, the convergence can be determined. The RELAX algorithm ends if this difference is less than a predefined threshold. Owing to the cyclic nature of the RELAX algorithm, the computational load is determined by several factors, such as the SNR, the data length, the number of padded zeros, the locations of the true line spectra, the details of how FFT is implemented, and the machine on which the program runs. On one hand, RELAX is more computationally efficient than many other spectra estimation algorithms involving heavy matrix manipulation, such as singular value decomposition and matrix inversion. On the other hand, RELAX is more computationally demanding than the conventional periodogram to trade complexity for performance. Works have been reported to further reduce the computation required by RELAX (Liu and Li, 1998). There are also other works reporting the computation complexity comparison between the RELAX algorithm and other competitive line spectra estimation algorithms (Shahid et al., 1999).

4.3.2.3 Examples The first example shows vital sign detection from the back of a human body with a carrier frequency of 5.8 GHz. Data were recorded when the subject was seating still in a chair. Both the periodogram and RELAX were applied to the measured data (shown in Fig. 4.26) with 257 samples and a sampling frequency of 20 Hz.

The RELAX algorithm was applied to the baseband signal, and a few intermediate stages are shown in Fig. 4.27. Because **B** is complex valued, both positive and negative frequency axes should be taken into consideration. In radar noncontact vital sign detection, $K = 5$ is capable of recovering the desired sinusoidal components when detecting from the front of the subject, and K was set to be 14 in this example to pursue higher accuracy.

Figure 4.27a shows that at the beginning of the RELAX algorithm, the strongest sinusoid with the {complex amplitude, frequency} of $\{-0.0924 + 0.0540i, -19.7726\}$ was directly estimated from the periodogram. This sinusoid was then subtracted from the periodogram, and the second strongest sinusoid was found to be $\{-0.0464 + 0.0768i, 20.0046\}$ in Fig. 4.27b. At that time, RELAX did not rush to estimate the third sinusoid, instead it tries to update these two already estimated sinusoids. Update starts by subtracting out the second strongest sinusoid $\{-0.0464 + 0.0768i, 20.0046\}$ from

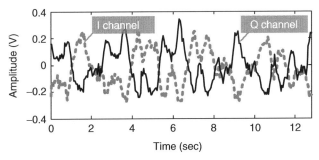

Figure 4.26 A set of baseband signal measured by a 5.8 GHz radar.

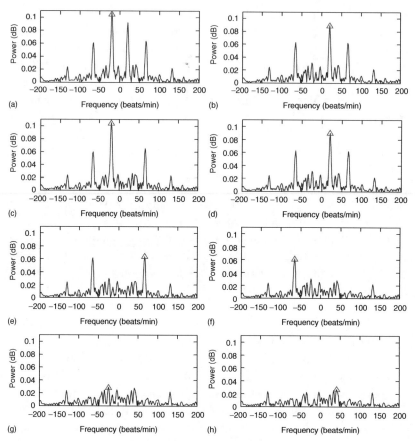

Figure 4.27 Intermediate stages when applying (a) RELAX for analyzing the baseband signal spectra. The estimates of the determined signals are updated by iterating, as (c) and (d) show. When it is the time to estimate a new signal, it will appear in spectra because RELAX removes all the estimated stronger sinusoidal component from the spectra, as (b), (e)–(h) show (Li et al., 2010a).

the original periodogram, then the sinusoid estimated in Fig. 4.27a was updated to be $\{-0.0901 + 0.0546i, -19.7915\}$, as shown in Fig. 4.27c. Then, in Fig. 4.27d this newly updated sinusoid was removed from the original periodogram, and the other sinusoid estimated in Fig. 4.27b was updated to be $\{-0.0464 + 0.0770i, 20.0066\}$. This mutual update repeats until the cost function between two successive iterations was less than $t = 10^{-3}$, indicating that both signals are good enough. By subtracting both of them from the original periodogram, the initial estimate for the third strongest sinusoid was found to be $\{-0.0342 + 0.0542i, 65.8566\}$ in Fig. 4.27e. It then began a new round of iterative updates of the three strongest sinusoids. This time, however, one sinusoid was updated by subtracting the other two from the periodogram. After convergence, by subtracting these three sinusoids from the periodogram, the fourth sinusoid was initially estimated to be $\{-0.0427 + 0.0440i, -65.7289\}$, as shown in Fig. 4.27f. Likewise, another round of iterative updates began and one sinusoid was updated by subtracting the other three from the periodogram. Figure 4.27g and h shows the initial estimates for the fifth and sixth sinusoids to be $\{-0.0027 - 0.0291i, -24.3334\}$ and $\{0.0233 - 0.0136i, 40.8858\}$, respectively. This procedure repeats until the parameters of all $K = 14$ sinusoids are estimated and updated.

As the true respiration and heartbeat rates have positive values, the consideration can be narrowed to the range of 0–120 beats/min. Figure 4.28 plots the estimates from RELAX in this range, together with the periodogram of the baseband signal. Five of the 14 sinusoids can be seen in this range. The respiration signal component at 20 beats/min was identified, while RELAX shows that the respiration consists two closely located sinusoids at 19.1 and 23.5 beats/min.

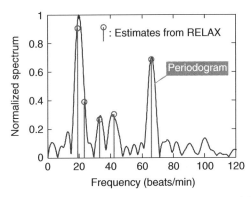

Figure 4.28 Final result of RELAX estimation to measured baseband signal when detecting from the back of the human body. Data length $N = 257$, convergence threshold $t = 10^{-3}$, $K = 14$ (Li et al., 2010a).

This was caused by the variation of respiration rate during the measurement period. The heartbeat signal component was identified as 65.9 beats/min.

To further demonstrate the advantage of RELAX, a computer-simulated baseband signal of extreme case, that is, the third-order harmonic of respiration signal located very close to the heartbeat signal, was used for processing. The ray-tracing model was used to simulate the Doppler radar vital sign detection.

The carrier frequency was set as 20 GHz, the baseband sampling period was set as 0.027 s, and the data length was 445, corresponding to a total sampling time of $445 \times 0.027 = 12.015$ s. Therefore the periodogram resolution limit is $1/12.015 = 0.0832$ Hz, which was converted to the "beats/min" unit of 4.992. The respiration rate was 21.5 beats/min, while the heart rate was set as 67.5 beats/min. The frequency difference between the heartbeat signal and the third-order harmonic of respiration signal was 3 beats/min that was less than the resolution limit. As a result, they form a single peak in between and periodogram cannot resolve either of them. On the other hand, RELAX algorithm successfully located the two frequency components as indicated in Fig. 4.29.

On the basis of the RELAX results, the estimates for the heartbeat and respiration rates were 21.5357 and 67.918 beats/min, respectively. And the third-order harmonic of respiration signal near the heartbeat

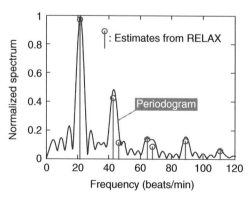

Figure 4.29 RELAX estimation to computer-generated baseband signal when detecting from the front of the human body. Data length $N = 445$, threshold $t = 10^{-3}$, $K = 14$ (Li et al., 2010a).

was found to be 64.6511, whose true value was 64.5 beats/min. It is shown that, although there exist two closely located sinusoidal components, RELAX still successfully resolves them and gains more insight from the spectra.

4.3.3   Joint Time–Frequency Signal Analysis

The electromagnetic energy backscattered from an unknown target can provide information useful for classifying and identifying the target. This is commonly accomplished by interpreting the radar echo in either the time or the frequency domain. In the time domain, scattering center analysis provides useful information for isolating the local features on the target. In the frequency domain, the singularity expansion method is widely used for extracting the target natural resonance frequencies, which contain information about the global features of the target. For target characteristics which are not immediately apparent in either the time or the frequency domain, sometimes the joint time–frequency representation of the radar echo can provide more insight into echo interpretation. In joint time–frequency analysis (JTFA), signals are analyzed in both the time domain and the frequency domain simultaneously.

A simple understanding of the JTFA fundamental can be described as follows. Although, frequency-domain representations such as the power spectrum of a signal often show useful information, the representations do not show how the frequency content of a signal evolves over time. For this task, JTFA can be used. JTFA is a set of transforms that maps a one-dimensional time-domain signal into a two-dimensional representation of energy versus time and frequency. There are numerous applications in both research and industry for JTFA. Examples include speech analysis, telecommunications, underwater acoustics, bioacoustics, geophysics, structural analysis, and radar detection.

There are a number of different transforms available for JTFA. Each transform type shows a different time–frequency representation. The short time Fourier Transform (STFT) is the simplest, and perhaps the easiest to compute, JTFA transform. For STFT, the well-known FFT is applied repeatedly to short segments of a signal at ever-later positions in time. The result can be displayed on a 3-D graph or a so-called 2-D

1/2 representation (the energy is mapped to light intensity or color values).

The STFT technique suffers from an inherent tradeoff between time resolution and frequency resolution, that is, increasing the first decreases the second, and vice versa. This coupling can skew the measurements derived from the transform, such as average instantaneous frequency. Other JTFA methods and transforms can yield a more precise estimate of the energy in a given frequency–time domain. Some options include: Gabor spectrogram, wavelet transform, Wigner–Ville distribution, and Cohen class transforms.

The Wigner–Ville distribution (Moghaddar and Walton, 1993) provides good localization of scattering mechanism. However, it introduces additional cross terms which lead to the so-called "ghosts" in the time–frequency plane. Originally developed by Goupillaud et al. (1984) to overcome the fixed resolution of the STFT, the theory of wavelets is attractive in various disciplines such as communication signal analysis, speech processing, and image compression. In Kim and Ling (1993), the wavelet transform was applied to represent electromagnetic backscattered data with better time–frequency characterization from complex targets. Different from the conventional STFT, which has fixed resolution in both time and frequency, the wavelet transform can provide variable resolution in time and multiresolution in frequency. The electromagnetic backscattered data can be elucidated from either the time domain or the frequency domain. In the frequency domain, the scattered signal comprises scattering mechanisms with widely different characteristic scales by taking advantage of multiscale windows. In the time domain, the variable time resolution property of the wavelet transform is ideal for providing good frequency resolution to isolate different natural resonances at different times of a radar echo.

There are many works on time–frequency analysis of radar signals for target detection. Ling et al. (1993) analyzed the backscattered signal from a coated strip with a gap in the coating. The results showed good localization of the different scattering mechanisms. It was found out that the time–frequency representation is well suited for identifying the dispersive surface wave contributions. In Trintinalia and Ling (1995), the scattering phenomenology in slotted waveguide structures is investigated using time–frequency processing of numerically simulated data. In Trintinalia and Ling (1997), an algorithm that combines inverse synthetic aperture radar (ISAR) processing with the joint time–frequency signal representation was developed as a means of extracting the nonpoint-scattering features from the standard ISAR image. The adaptive joint time–frequency ISAR algorithm has been successfully tested using data generated by the moment-method simulation of simple structures and the chamber measurement data from a scaled model airplane. The results showed that nonpoint-scattering mechanisms can be completely removed from the original ISAR image, leading to a cleaned image containing only physically meaningful scattering centers. When displayed in the frequency-aspect plane, the nonpoint-scattering mechanisms can be used to identify target resonances and cutoff phenomena. More details of joint time–frequency analysis for radar signal processing can be found in Chen and Ling (1999).

5

APPLICATIONS AND FUTURE TRENDS

Microwave noncontact motion sensing technologies introduced in previous chapters have a variety of potential applications. The ability of penetrating through rubbles to detect human respiration and heartbeat led to the initial application and research effort of searching earthquake survivors. Recent research efforts have been focusing on developing compact, lightweight, low power systems for portable, handheld, and embedded applications. The advances in wireless technologies and semiconductor technologies have helped the miniaturization of radar system hardware, which paves the way for broader range of applications. Among many potential applications, healthcare monitoring, whether in hospital or home, seems to be grabbing most of the attention. In addition to monitoring human patients, the noncontact motion sensing technologies can also be applied to monitor animals in veterinary care. Besides healthcare and veterinary care, the ability to perform noncontact, covert detection of respiration, heartbeat, as well as other body motions enables other applications such as security monitoring, battlefield triage, and

Microwave Noncontact Motion Sensing and Analysis, First Edition.
Changzhi Li and Jenshan Lin.
© 2014 John Wiley & Sons, Inc. Published 2014 by John Wiley & Sons, Inc.

law enforcement. Furthermore, the miniaturized micro radar can potentially be embedded in other equipment such as smart phones and laptop computers for personal use. In this chapter, several examples of potential applications are described. Future trends and industry outlook are discussed to conclude this book.

5.1 APPLICATION CASE STUDIES

5.1.1 Assisted Living and Smart Homes

Owing to the increase in the aging population and target on energy efficiency, the need for assistive living and smart home technologies are arising rapidly. Microwave motion sensing technologies have the potential to revolutionize home healthcare delivery and energy efficiency by estimating activity level in real time.

According to the United Nations in 2005, 20% of the world's population was aged 60 years or over. It is estimated that by 2050 that proportion will increase to 32%. This aging trend has major implications on the future of healthcare delivery, policy and quality of life for people around the world. With the growing emphasis on the adoption and impact of health-related information technology, researchers and practitioners are increasingly focusing on assistive living and smart home technologies. Microwave radar motion sensing has a large potential to advance wireless sensing in such applications. A sensor platform capable of detecting a person's vital signs and activity levels is poised to make a strong impact. Because the platform can be deployed in a variety of locations (home, hospital, emergency room, office space, etc.), it can update a person's health condition and activity behavior pattern in real time. Studies have been reported for gait and activity analysis among elderly subjects. Hynes et al. (2011) introduced a system solely utilizing accelerometers in cellular handsets to remotely monitor activity characteristics of elderly patients at home. The system provided one potential methodology to monitor and analyze a patient's daily activity characteristics and present historical data collected over time. However, accelerometers have limited resolution and need to be attached to the person being monitored.

In Reyes et al. (2012), a monitoring system, named as *VitalTrack*, was realized using a motion sensing radar capable of determining a person's activity levels. A 5.8 GHz homodyne quadrature Doppler radar is used to conduct experiments and measure signals. The I/Q signals are combined and a moving average of the biosensor data collected can be obtained by taking the average of the first subset of n data points. The moving average $A(t)$ is defined as

$$A(t)_{I/Q} = \frac{[B(t)_{I/Q_1}^2 + \cdots + B(t)_{I/Q_n}^2]}{n}. \qquad (5.1)$$

The parameter n is determined experimentally based on the responsiveness of the radar sensor. The best approximation for n, considering the responsiveness and sensitivity of the sensor to a variety of activity levels, is $25-100$. The averaging process is repeated over the entire data series as data is collected from the monitoring platform.

The overall block diagram of the platform is shown in Fig. 5.1. The system consists of distinct components: a monitoring platform mounted from the ceiling or observation point within the home or living space, a laptop base station to collect and process data from the platform, and web services capabilities to enable remote monitoring of the overall system from any mobile device or computer connected to the Internet.

The mobile sensing platform was constructed using a 0.220 in. thick transparent Plexiglas. A two-level platform was designed using

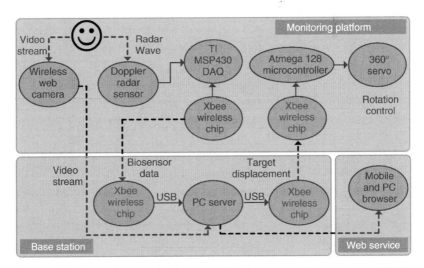

Figure 5.1 Overall VitalTrack system block diagram (Reyes et al., 2012).

a 17 in. × 11 in. piece for the top level and a 10 in. × 8 in. piece for the bottom level. The bottom and top levels of Plexiglas are connected using aluminum brackets and the following components are mounted on the lower level of the platform: TI MSP430 microprocessor and XBee transmitter board, a wireless web camera, a second XBee transmitter board, MAVRIC-IIB microcontroller board, 1 × 6 V battery, on/off switch, and the GWS S125 3T servo. The platform is mounted from the ceiling to provide a 360° observation area. A rotating servo attached to the pan/tilt kit allows the radar antennas and camera to move as a person is detected and tracked.

The Doppler radar sensor measurements can be obtained by targeting the person's chest wall. However, it is impossible to assume that a person is within the sensor and platform's observation range at all times. Hence, a web camera and custom face detection algorithms using OpenCV libraries were adopted to detect and track a person as their motion is captured. By using detection and tracking, the Doppler radar sensor platform is able to follow the person in their living environment, realign the sensor with the person's chest wall, and thus provide accurate and continuous monitoring of health and activity levels. A C++ application was developed on the laptop base station to track the person's face within the web camera frame consisting of 640 × 480 pixels. As the person moves and their location relative to the center of the frame changes, the system is able to detect if the person has moved to the right or left of the platform. The system uses this information to rotate the platform along its axis, keeping the person in its view at all times. The low power 802.15.4 Zigbee protocol for wireless communication is used for data transmission from the monitoring platform to the base station.

Figure 5.2 shows a typical office space layout with an overlay visualization of activity-level results. Concentric circles indicate activity levels. Sensing using a Doppler radar sensor in a static location is highly directional. Through the use of a rotating platform attached to the ceiling with computer vision capabilities, the system can track a person's movements in an omnidirectional manner.

Measurements have been performed to verify the functionality of the platform. In the first measurement, the radar sensor was programmed to run continuously for a period of six hours beginning from 12 AM until 6 AM. The experiment was used to establish a ground truth

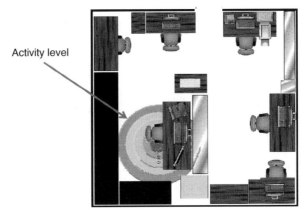

Figure 5.2 Layout of office space with overlay of activity-level visualization (Reyes et al., 2012).

measurement and collect data to represent a state of inactivity. "Inactivity" is defined as the state in which no activity levels are detected either because the person is completely stationary or the person is not present in the observation range of the platform. Output from the face detection application provides confirmation that a person is not detected. The Doppler radar sensor was mounted in an empty office space, which would allow for a period of inactivity to be detected. The peaks observed at the beginning and end of the measurement period are spontaneous activity levels captured as the researcher exits and enters the space being monitored. Figure 5.3a shows the resulting data detected for a state of inactivity with 0 V detected. The radar data collected demonstrates that a state of inactivity was observed for approximately six hours and any noise in the system is the cause of small spikes observed. Video recorded from the web camera verifies that no facial detection or activity occurred in the frame of the sensor during these six hours.

The second measurement identified the difference between spontaneous activities (which occur for a brief instant in time) and sustained activities (maintained continuously for a period greater than 30 s). For example, walking past the platform or abrupt motion of the chest wall would be considered spontaneous activities with varying levels of low to high activity. A sustained activity is maintained continuously and is

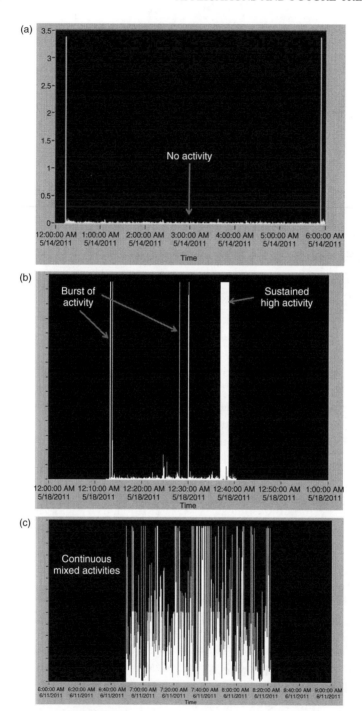

Figure 5.3 (a) Inactivity observed; (b) spontaneous and sustained activity detected; (c) continuous mixed activity levels (Reyes et al., 2012).

generally captured as high activity. For example, exercising or continuous body movements in the observation range will lead to sustained activity levels captured.

Figure 5.3b shows the resulting data captured for a period of 30 min. The width of the spikes observed is used to determine whether the activity is spontaneous or sustained. Spontaneous activity levels are detected twice near 12:15 AM, once at 12:28 AM and again at 12:32 AM. In this case, the sustained activities tested were similar to Fig. 5.3a walking past the sensor in the observation window of the platform. As the person walks past the sensor, a spike occurs and it is captured as a spontaneous activity that only occurs for a brief number of seconds. The sustained activity tested was a 5-min period of stretching and jogging in place. This period of sustained activity (greater than 30 s) is captured between 12:37 AM and 12:42 AM. The larger width of the captured spike in activity level indicates that the resulting activity is a sustained one. Some noise is detected around 12:26 AM and is due to noise in the environment and possibly movement or slight vibrations from the wall or surrounding objects.

The third measurement consisted of monitoring a person sitting at their office desk over a period of approximately 2 h. The Doppler radar sensor platform was placed at 0.5 m away from the person with direct line of sight to the person's chest wall. During the observation period, the target was asked to perform spontaneous activities such as standing up, deep breathing, and walking in and out of the monitoring area. These events are randomly recorded as medium to high level activities. By averaging the total outputs at the end of the day, it would be possible to estimate a person's lifestyle and activity levels, which could be correlated with other sensory data to determine overall health. Figure 5.3c shows the combined plot of activity levels. The measurement resolution can be improved if detection distance is reduced to improve the signal-to-noise ratio.

On the basis of the experiment result, the VitalTrack system enables continuous noncontact and nonintrusive monitoring of a person's vital signs (at rest) and activity levels (in motion). The experimental results indicate that the Doppler radar is able to distinguish between spontaneous and sustained activity levels. It is expected that the Doppler radar technique for activity-level monitoring would serve as a preferred and effective alternative to monitoring daily activity for both healthcare and energy efficiency applications.

5.1.2 Sleep Apnea Diagnosis

Sleep apnea, which is a sleep disorder characterized by pauses in breathing during sleep, is a common respiratory-related syndrome among many adults in the United States. According to the National Institutes of Health (National Institute of Health News), sleep apnea affects more than 12 million Americans. Because of the lack of awareness by the public and health care professionals, the vast majority remain undiagnosed and therefore untreated, despite the fact that this serious disorder can have significant consequences. Untreated sleep apnea can cause hypertension and other cardiovascular disease, memory problems, cognitive impairment, psychological distress, weight gain, impotency, excessive daytime sleepiness, and headaches. Moreover, untreated sleep apnea may be responsible for job impairment and motor vehicle crashes.

Traditionally, the diagnosis and treatment of sleep apnea require polysomnography (PSG) in sophisticated sleep labs, which is expensive and severely lacking in many places. Even with a sleep lab, most of the monitoring technologies are based on wired probes or chest straps, which limit the activity and create discomfort to patients being monitored. Therefore, long-term monitoring of vital signs without affecting normal life in a comfortable environment is difficult. Regarding the efficacy and performance, traditional wired technologies such as electrocardiography (ECG) is not a perfect solution. The performance of such methods relies on contact technologies that are likely to be disturbed by activities of monitored subjects, making it difficult for accurate continuous monitoring. Also, the surface-loading effect and body fluid can reduce the accuracy of probe-based measurement. As a result, current laboratory PSG is cumbersome, inconvenient, and expensive, causing considerable interest in portable or noncontact monitoring of the condition.

The radar motion sensing technology offers a noncontact alternative to traditional wired methods. The noncontact approach provides benefits for easy implementation and less constraint on human activity. It is robust against body fluid and does not involve concern of radiation safety because the transmit power is very low. The main challenge for robust long-term monitoring of sleep apnea is whether the radar can still detect vital signs when patient changes sleep positions (i.e., fetus, log, yearner, soldier, freefall, and starfish) in an unconfined environment. To demonstrate the feasibility of radar-based monitoring, experiments were carried out in 2005 at the University of Florida to monitor vital signs from four sides of a human body, that is, front, back, left, and right.

A Ka-band heartbeat detector was used in a lab environment. The test setup, viewed from the top, is shown in Fig. 5.4. The subject, breathing normally, was seated at a distance away from the antenna. A wired fingertip pulse sensor (UFI_1010 pulse transducer) was attached to the index finger during the measurement to provide the reference heartbeat.

During experiment, the radar-measured heartbeat rate (HR) was evaluated by "heart rate accuracy," which is calculated as the percentage of time the calculated rate is within 2% of the reference rate. The experiment conditions were designed as combinations of the following parameters: two power levels of 350 and 14.2 μW; five different distances from the antenna: 0.5, 1, 1.5, 2, and 2.5 m; and measuring from four sides of the body. The measured results of HR accuracy for all the above combinations are listed in Table 5.1.

The experiment indicates that the detection accuracy from any side of the body is better than 80%. In addition, the measurement from the back shows the best performance, which will be discussed later. The results also indicate that better accuracy can be achieved with higher power. Measurement results also show, to some surprise, that the HR accuracy detected from the back is better than that detected from other sides. By analyzing the spectra, an apparent difference between the

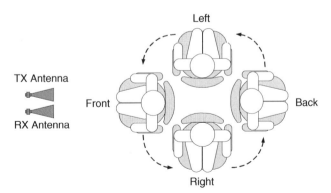

Figure 5.4 Top view of the test setup to measure vital signs from four sides of a human body.

TABLE 5.1   Summary of Heart Rate Detection Accuracy[a]

Distance, m	Front, %	Left, %	Right, %	Back, %
		14.2 μW		
0.5	99.1	96.3	100	97.6
1	89.8	89.8	93.2	100
1.5	98.9	89	93.8	94.3
2	85.2	80.5	97.4	93.6
2.5	83.3	85.7	85.1	85.5
		350 μW		
0.5	100	100	100	100
1	94.8	94.7	93.2	100
1.5	98.1	97.6	100	100
2	100	100	100	100
2.5	95.1	100	95.2	97.2

[a]Adapted from Li et al. (2006a).

spectra measured from the front and from the back can be discovered as shown in Fig. 5.5. The normalized baseband output spectra were detected from the front and the back of the body at 2.0 m distance and under the power level of 350 μW. It shows stronger harmonics of the respiration signal when detecting from the front than from the back. This is because the physiological movement due to respiration is much stronger for the front chest wall than the back. Therefore, the difference in respiration harmonic interference caused the difference in detection accuracy.

The above experiments have successfully demonstrated the feasibility of using radar to monitor vital signs from four sides of a human body. On the basis of the observation that higher accuracy can be obtained by measuring from the back of a human body, a system shown in Fig. 5.6 was set up for overnight monitoring of vital signs during sleep. A fingertip sensor with wired connection to signal recording unit was used to provide a reference HR. As a drift in frequency or a burst of phase noise may change the detection point from an optimum point to a null point for the Ka-band double-sideband transmission radar, the high frequency yttrium–iron–garnet (YIG)-tuned oscillator in the radar was replaced with an external signal generator during the overnight measurement.

The experiments conducted were taken from midnight to the next morning. It is unavoidable that sometimes the subject under test may

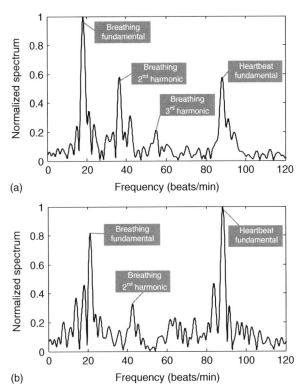

Figure 5.5 Normalized spectrum comparison at 2.0 m distance from the front (a) and the back (b) with radar transmitted power lever of 350 μW (Li et al., 2006a).

move his/her body during measurement. Sometimes, the subject may have motion on the fingertip without moving the body, in which case the noncontact detector is more likely to get an accurate rate than the fingertip measurement. Figure 5.7a shows a 1-min record of such a case during a 6-h monitoring. On the other hand, the subject may also move the body without disturbing the fingertip sensor, making the fingertip referenced rate more reliable, as shown in Fig. 5.7b. Combining these factors, the fingertip sensor, although still used as a reference, is no longer the standard for judging the accuracy of the noncontact radar detector. Therefore, we refer the percentage of time the detected rate is within 2% of the reference rate as the *degree of agreement* of the two detection methods.

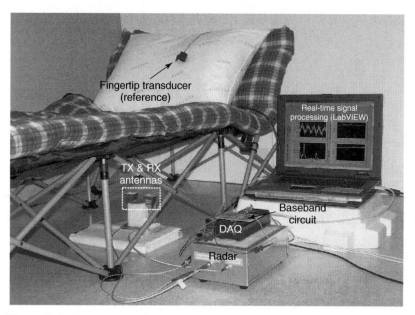

Figure 5.6 Measurement setup for overnight monitoring of respiration and heartbeat rates.

For 6 h of noninterrupted monitoring, the degree of agreement between the noncontact radar sensor and the fingertip sensor can achieve as high as 90%. Figure 5.8 shows a 6-h overnight measurement record of detected HR and respiration rate (RR) using noncontact radar detector, as well as the detected HR using the wired fingertip sensor. In plotting these results, all the rates were running averaged using a window size of 60 s. This is aimed to observe the long-term trend in rate change, and to eliminate the apparent false detections of both detected rates caused by body movement. Before the running average, the degree of agreement between the two monitoring methods is 89.33%. By using running average with a window size of 10 s to eliminate apparent false detections, the degree of agreement achieves 92.96%.

Therefore, the results show that the radar motion sensor is able to monitor human physiological movements from all four sides of a human body in a long term. Radar noncontact motion sensing is an attractive alternative technology, with advantages of easy implementation, noncontact, low power, no constraint to normal life activities,

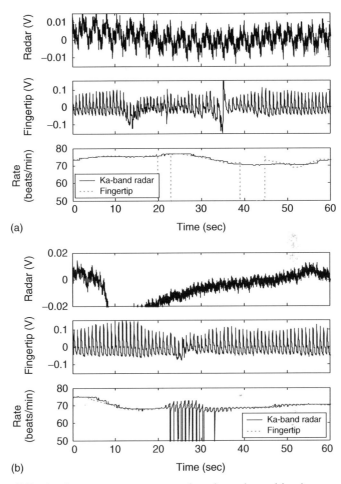

Figure 5.7 1-min measurement records when the subject's movements caused interruptions: (a) the noncontact radar was detecting normally while the fingertip sensor data shows interruptions; (b) the fingertip sensor works normally while the noncontact radar data shows interruptions (Li et al., 2006b).

and robust against body fluid and artifacts, for the applications of overnight monitoring of sleep apnea syndrome.

5.1.3 Wireless Infant Monitor

Traditionally, ECG and medical technologies used to monitor vital sign information such as heartbeat, respiration, and blood pressure are often

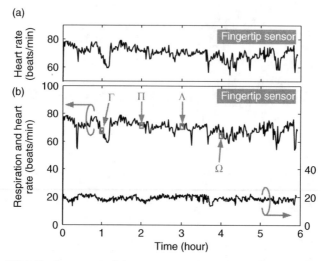

Figure 5.8 Six-hour overnight measurement record of (a) detected heart rate using wired fingertip sensor; (b) detected heart and respiration rates using noncontact radar detector.

wired. The performance of such conventional methods relies on contact technologies that are likely to be disturbed by activities of monitored subjects (particularly infants and newborns), making it difficult for accurate continuous monitoring. Also, the surface-loading effect and body fluid on the small body figure of infants/newborns can reduce the accuracy of probe-based measurement. From application point of view, it is beneficial to have a noncontact technology that can monitor vital signs for early action to Sudden Infant Death Syndrome (SIDS, the third leading cause of infant mortality) of infants. SIDS is the unexpected, sudden death of a child under age one in which an autopsy does not show an explainable cause of death. Most of SIDS deaths occur without any warning or symptoms when the infant is thought to be sleeping. Therefore it is crucial to monitor the vital sign of infants younger than 6 months, especially for newborns. Moreover, infants can also have high blood pressure that is usually caused by prematurity or problems with the kidneys or heart. Untreated high blood pressure can eventually lead to damage to the heart, brain, kidneys, and eyes. But if it is monitored and treated, an infant can lead an active normal life. Therefore, a low cost, noncontact technology optimized for long-term monitoring can greatly reduce neonatal mortality and morbidity.

The microwave motion sensing technology makes it possible to monitor the heartbeat, respiration, and blood pressure of infants and newborns without any probe attached to the body. Such a device may display and store vital sign data, and send the data to a remote server, if required. In addition, the device can set off alarms when vital signs of the infants/newborns fall in the alert range.

On the basis of the Doppler phase modulation effect, such a device senses the physiological movement with high sensitivity while avoiding false alarm by the random body movement cancellation algorithm discussed in Chapter 4. Physiological movements reflect respiration, heartbeat, and pulse wave. Pulse wave velocity (PWV) has been demonstrated as an independent predictor of diseases such as hypertension, vascular damages, diabetics with presumed athero-sclerosis, and end-stage renal failures (Asmar, 1999). By measuring the pulse waveforms at two body locations (e.g., carotid-femoral and brachial-ankle) of a newborn, PWV can be obtained, making it possible to obtain blood pressure information.

Figure 5.9 shows a setup for infant monitoring. Electronic beam steering radar is used to monitor the physiological movements and pulse movements at the heart and the calf of infants. The physiological movements reflect respiration and heartbeat in a straightforward fashion. In the meantime, the nonlinear phase demodulation obtains

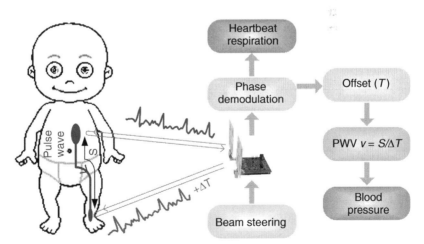

Figure 5.9 Noncontact monitoring of respiration, heartbeat, and pulse wave velocity of a baby.

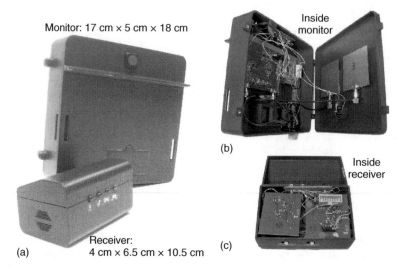

Figure 5.10 Exterior of a prototype system designed and fabricated (a), interior of the monitor unit (b) and interior of the receiver unit (c).

the pulse waveforms at the two locations. The phase difference ΔT of a pulse wave at different locations of the body will be obtained. On the basis of the estimated distance S between the two locations, the mean PWV can be calculated as $v = S/\Delta T$. Artery blood pressure will be indirectly obtained from the PWV.

Figure 5.10 shows the exterior and interior of a prototype baby monitor system developed at the University of Florida (Li et al., 2009a). To provide flexibility and ease of use, both the receiver and monitor units can be powered by batteries or plugged directly into wall power. The receiver unit can travel up to 50 m away from the monitor while still being able to alert the parent. For operating range larger than 50 m, the receiver needs to be connected to a laptop, which forwards the alarm to parents' cell phone or personal digital assistant (PDA). The cost for mass production of the monitoring system is targeted to be under $80 for the monitor and receiver pair; multiple receivers could be added to work with a single monitor. Limited by the package facilities available, the prototype has a relatively large size. With more refined packaging, the receiver could be reduced to the size of an ordinary cell phone. The monitor could also be significantly reduced, with the limiting factor being the antennas used.

The fear of SIDS and the prevalence of infant breathing problems have many parents searching for a way to monitor the health and well-being of their child, especially, while sleeping unattended. Therefore, it is expected to have a prime market of first time parents with the income and the inclination to buy baby monitors that provide more than just sound or video. A noncontact heartbeat/respiration monitor would remotely watch over the health of the child and provide parents with peace of mind.

5.1.4 Measurement of Rotational Movement

Most of the applications discussed so far have been limited to the detection of one-dimensional movement. Some industrial applications require the detection of rotational movement. For example, it is desirable to monitor the spin speed of motors and servos in macroscale machineries and microscale microelectromechanical system (MEMS) devices. Therefore, whether the Doppler radar sensing mechanism could be extended to the detection of angular speed is an interesting topic.

When Doppler radar is used to detect one-dimensional and one-way movement, the movement $x(t) = v_0 t$ produces the well-known *Doppler frequency shift* to signal reflected from the object. As a result, the baseband signal $B(t)$ will have a single frequency equal to $2v_0/\lambda$. The speed of the moving object can be determined by measuring the radar baseband output frequency. On the other hand, when the Doppler radar is used to detect one-dimensional periodic movement, the Doppler phase modulation effect will produce a series of frequency components, with the fundamental frequency equal to the periodic movement frequency. Therefore, the movement frequency can be determined by measuring the fundamental tone at the radar baseband output. Moreover, the amplitude of periodic movement can be determined based on the relative strength among the harmonics of radar baseband output, providing a low cost measurement method without requirement on signal strength calibration.

Figure 5.11a and b show the scenario of radar noncontact monitoring of rotational movements. The targets are placed at a fixed position in front of the radar, while they are rotating at a high speed driven by a motor. In these cases, the objects have rotational movement instead of one-dimensional position change. As a result, the objects can no

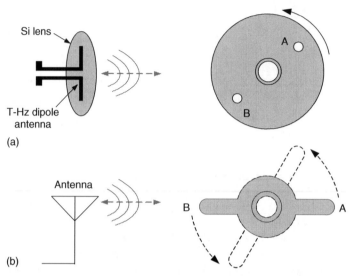

Figure 5.11 Radar noncontact monitoring of rotational movements. (a) Monitoring of a microscale rotating MEMS silicon disk using Terahertz carrier frequency. (b) Monitoring of a macroscale rotating motor.

longer be simplified as a single point; otherwise there will be no variation in the reflected EM wave. It is straightforward to understand that the radar cannot detect rotational movement of a homogeneous object because a homogeneous rotating target will produce no change to the reflected EM wave.

In order to make the Doppler radar noncontact detection possible for a rotating silicon disk as shown in Fig. 5.11a, it is desirable to implant some structure, for example, one or two silicon-dioxide bubbles, to change the reflected EM wave as the disk rotates. However, if there is only one bubble (e.g., bubble A) implanted, it will be difficult to balance the mechanical rotation. Therefore, silicon-dioxide bubbles A and B are implanted in symmetry to the center rotating axle. In Fig. 5.11b, the two arms (A and B) of the rotating object change the reflected EM wave as their orientation changes. As a result, the radar will receive a time-varying signal that changes periodically as the object rotates.

The time-varying reflected EM wave produces a modulated received signal for the Doppler radar. If the received signal can be down-converted to baseband, it is possible to recover the information of angular speed. Two numerical simulations based on the ray-tracing model will be performed to demonstrate the feasibility of Doppler radar for noncontact detection of the above rotational movements. In order to demonstrate both micro- and macro-scale detections, the target in Fig. 5.11a was assumed to be a MEMS rotating silicon disk with a radius of 60 μm, while Fig. 5.11b was assumed to be a rotating object with two 3 cm arms.

For the detection of MEMS rotating silicon disk, a hypothetic Doppler radar operating at 1 THz is used so that the wavelength can provide enough sensitivity to small movements. In Terahertz frequency range, it is possible to integrate the antenna and lens, generating a Terahertz beam with good directivity. On the basis of reported Terahertz transmission architecture (Dragoman and Dragoman, 2004), the Terahertz signal source in Fig. 5.11a is connected to a coplanar stripline (CPS) terminated with a dipole antenna metallized on the semiconductor substrate and having two arms with lengths of about 40 μm. A Si spherical lens is mounted above the antenna to collimate the emitted Terahertz radiation. In the simulation study performed by Li and Lin (2009b), the silicon disk driven by MEMS motor is placed in front of the Terahertz radar, having an angular speed of $\omega_0 = 2\pi f_0$, where $f_0 = 600$ RPM (rotations per minute). As discussed above, two silicon-dioxide bubbles A and B placed in symmetry to the center axle are implanted into the disk to keep mechanical balance.

In simulation, A and B are simplified into two points rotating around the motor axle. At each time step, the location of A and B are calculated. Then the strength of EM wave reflected from the two points is decided based on the radiation pattern and the two points' distance from the signal source. The received signal strength is obtained and recorded by adding the strength of reflected EM waves. Finally, the spectrum of baseband signal can be obtained.

Figure 5.12 shows the detected baseband spectrum. For each silicon-dioxide bubble (either A or B), the reflected EM pattern has a period that is the same as the period of the mechanical rotation. Therefore, the fundamental baseband frequency tone should be 600 RPM when the signal is only produced by either A or B, which is shown in Fig. 5.12a. However, when two bubbles are implanted in symmetry to the center axle, the reflected EM pattern will repeat every half rotation cycle, as the received signal will be the same

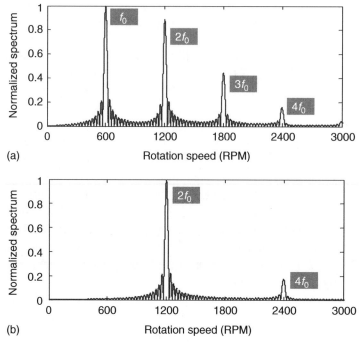

Figure 5.12   Detected baseband spectrum. (a) When a single SiO_2 bubble A was implanted into the Si disk; (b) when SiO_2 bubble A and B are implanted symmetrically into the Si disk.

after the disk rotates $180°$ (A and B interchange their positions). As a result, the baseband fundamental frequency detected with A and B implanted into the disk will be twice the rotation frequency of f_0, as shown in Fig. 5.12b. This "frequency doubling" effect is different from the measurement of one-dimensional periodic movement.

In the case of macroscale mechanic rotation as shown in Fig. 5.11d, the rotating object is not omnidirectional. Therefore, no extra procedure is needed to modify the EM property of the target. The total length of the two arms is 60 mm. The angular speed is $\omega_0 = 2\pi \times 600$ RPM. The radar has a carrier frequency of 5 GHz, detecting the movement from 1 m away. A 5×5 antenna array of omnidirectional antennas spaced by $\lambda/2$ is used in the simulation to transmit and receive RF signals. Because different parts on the arms of the object produce different reflected EM strength, the ray-tracing model

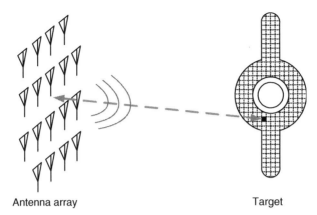

Antenna array   Target

Figure 5.13 Generating of the mesh grid for ray-tracing modeling of radar noncontact measurement of rotation speed.

given in Chapter 4 is used to take into account the change of signal amplitude and phase reflected from different parts.

As shown in Fig. 5.13, the object is partitioned using a fine mesh grid. The signal reflected from each cell is calculated based on the antenna array radiation pattern and the distance from the cell to the antenna array. The total received signal is obtained by integrating the signal reflected from each cell. A residue phase of $90°$ is assumed in the simulation. Fig. 5.14a shows the simulated time domain baseband output in the two quadrature channels. Figure 5.14b shows the radar output baseband spectrum obtained from complex signal demodulation.

Similar to the case of MEMS disk, the baseband output has a fundamental frequency that is twice the object's rotation speed. Because the carrier wavelength is comparable to the object size, strong harmonics due to nonlinear phase modulation are observed at the baseband output.

The above simulations demonstrate that the radar noncontact motion detection technology can be extended from detecting one-dimensional movement to detecting two-dimensional rotational movement. It has been shown that in noncontact rotational movement monitoring, the baseband output has a fundamental frequency that is multiple times of the frequency of the rotational movement. The simulations cover the detection in macroscale rotational movement as well as microscale

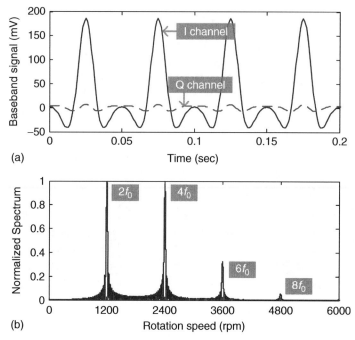

Figure 5.14 Baseband time domain signal (a) and complex signal demodulated spectrum (b) detected with a carrier frequency of 5 GHz and a residue phase of 90°.

MEMS rotation, adding a lot of new applications to the Doppler radar motion sensing technology.

5.1.5 Battlefield Triage and Enemy Detection

Battlefield triage is the process of determining the priority of injured soldiers' treatments based on the severity of their condition. It rations treatment efficiently when resources are insufficient for all to be treated immediately. Triage may result in determining the order and priority of emergency treatment, the order and priority of emergency transport, or the transport destination for the injured soldiers. Triage may also be used for patients arriving at the emergency department, or to telephone medical advice systems, among others. As the medics have a very high casualty rate, it is therefore a vital importance to prioritize which soldiers to attend to first. The first step is to detect life signs—if a soldier is dead or alive, and prioritize recovery of live soldiers. The second step is to obtain vital signs from live soldiers, and use this to prioritize which are in most urgent need of attention (Boric-Lubecke et al., 2008).

The noncontact motion sensing technology provides the means to detect life signs, respiration, and/or heartbeat, at a distance, even for subjects lying motionless, for example, unconscious subjects, wearing body armor, and hidden from direction view. As this technology can deliver HR information with high accuracy, it may also enable the assessment of a subjects' physiological and psychological state based on heart rate variability (HRV) analysis. Thus, the degree of a subject's injury may also be determined. The software and hardware challenges for life sign detection and monitoring for battlefield triage include heart signal detection from all four sides of a human body, detection in the presence of body armor, the feasibility of HRV parameter extraction, and the assessment of the degree of injury and thus urgency of intervention. Li et al., (2006a) has demonstrated that Doppler radar cardiopulmonary monitoring is possible from the four sides of a human body, and even in the presence of body armor that radio waves clearly cannot penetrate. Preliminary data indicated that HRV parameters may be extracted from such measurements, opening the door to remote medical diagnostics.

On the other hand, microwave motion sensor can be used to detect enemy in battlefield. By sensing respiratory and heartbeat motion of a human body, this technology can detect hiding enemies through the walls and obstacles. The challenges in this application include sensitivity of the radar and noise produced by movement of the radar itself. The sensitivity improvement techniques in Chapter 4 can be used to improve the performance. If the radar is handheld by a soldier, then shaking noise cancellation techniques are also necessary. Examples of shaking noise cancellation method have been discussed in Chapter 4.

5.1.6 Earthquake and Fire Emergency Search and Rescue

By detecting both the body motion and the physiological motion of victims buried under earthquake rubbles or in burned buildings, the sensing technology can improve the efficiency of search and rescue. The challenge is for EM wave to effectively penetrate rubbles and complex structures.

In the 1980s, a microwave life-detection system, which operates at the X-band (10 GHz) for sensing the physiological status of soldiers lying on the ground of a battlefield, was reported (Chen et al., 1986). It turned out, though, that such an X-band microwave beam cannot penetrate earthquake rubble or collapsed building debris deep enough to locate buried human victims. For this reason, two other systems, one operating at 450 MHz and the other at 1150 MHz (Chen et al., 2000a), have been implemented. On the basis of a series of experiments, it was found that an EM wave of 1150 MHz can penetrate earthquake rubble (layers of reinforced concrete slabs) with metallic wire mesh easier than that of 450 MHz. However, an EM wave of 450 MHz may penetrate deeper into rubble without metallic wire mesh than that of 1150 MHz. A key technology implemented in these works is a microprocessor-controlled clutter-cancellation system, which creates an optimal signal to cancel the clutter reflection from the rubble and the background. In this technique, a phase shifter and an attenuator were digitally controlled by a microcontroller to provide a delayed version of the transmitted signal. The delayed version of the transmitted signal was combined with the received signal using a directional coupler. Additionally, a microwave power detector was used to monitor the DC level of the combined signal, serving as the indicator for the degree of the clutter cancellation. The microcontroller automatically adjusts the phase delay and attenuation to minimize the DC level in the combined signal. And an optimal setting for the clutter cancellation corresponds to the point where the DC level of combined signal was minimized.

In practice, minimization of the device size and prolonging of battery life for portable application are two targets for commercialization developments. Low power CMOS system-on-chip integration is thus desirable.

5.1.7 Tumor Tracking in Radiation Therapy

Cancer is a leading cause of deaths over the world. According to the World Health Organization, global cancer rates could increase by 50% to 15 million by 2020. Radiation therapy is a major modality for treating cancer patients. Studies have shown that an increased radiation dose to the tumor will lead to improved local control and survival rates. However, in many anatomic sites (e.g., lung and liver), the tumors can move significantly ($\sim$2–3 cm) with respiration. The respiratory tumor motion has been a major challenge in radiotherapy to deliver sufficient radiation dose without causing secondary cancer or severe radiation damage to the surrounding healthy tissue (Jiang, 2006a; Jiang, 2006b). A typical example is the treatment of lung cancer that is the leading cause of cancer deaths in high income countries. Every day, approximately 439 Americans die from lung cancer. In fact, more people die from lung cancer each year than breast, prostate, colon, liver, kidney, and melanoma cancers combined. The treatment outcome of the current modalities has been poor; the 5-year overall survival rate for lung cancer is about 15%. Radiotherapy is a major modality for treating patients with lung cancer, either alone or combined with surgery and chemotherapy. Because lung tumors can move significantly with respiratory motion, it is very difficult to deliver sufficient radiation dose without damaging the surrounding healthy lung tissue and causing severe side effects such as pneumonitis.

Motion-adaptive radiotherapy explicitly accounts for and tackles the issue of tumor motion during radiation dose delivery, in which respiratory gating and tumor tracking are two promising approaches. Respiratory gating limits radiation exposure to a portion of the breathing cycle when the tumor is in a predefined gating window. Tumor tracking, on the other hand, allows continuous radiation dose delivery by dynamically adjusting the radiation beam so that it follows the real-time tumor movement. For either technique to be effective, accurate measurement of the respiration signal is required. Conventional methods for respiration measurement are undesirable because they are either invasive or do not have sufficient accuracy. For instance, measurement based on fiducial markers requires an invasive implantation procedure and involves serious risks to the patient, e.g., pneumothorax for lung cancer patients (Laurent et al., 2000). On the other hand, measurement of external respiration surrogates using infrared reflective marker, spirometer, or pressure belt generally lacks sufficient accuracy to infer the internal tumor position, because they only provide a point measurement or a numerical index of respiration. In addition, these devices have to be in close contact with the patient in order to function (Li et al., 2011). This often brings discomfort to the patient and can lead to additional patient motion during dose delivery. To that end, accurate respiration measurement that does not require invasive procedures or patient contact is urgently needed in order to realize the potential of motion-adaptive radiotherapy.

CW radar sensor provides a noncontact and noninvasive approach for respiration measurement. Instead of measuring the marker, it directly measures the periodic motion of the body, which has better correlation with lung tumor motion. Moreover, the radar system is insensitive to clothing and chest hair, because of microwave penetration, making it better than the existing contact devices that are sensitive to the surrounding environment. In radar respiration measurement, the radar sensor suffers from DC offset at the RF front-end output, which is mainly caused by the reflections from stationary objects surrounding the body. The DC offset may saturate or limit the dynamic range of the following stages of baseband amplifiers. To overcome this demerit, AC coupling has been commonly used in radar sensors. However, because of the high pass characteristics of the coupling capacitor, AC coupling leads to significant signal distortion when the target motion has a very low frequency or a DC component. Respiration is such a motion with a low frequency of less than 0.5 Hz, and tends to rest for a while at the end of expiration (EOE), that is, there is a short stationary moment after lung deflation. This is a problem in radar respiration measurement. To deal with it, DC-coupled radar discussed in Chapter 4 was adopted (Gu et al., 2012a).

Figure 5.15 shows the motion-adaptive radiotherapy system based on radar respiration sensing. The radiotherapy process includes two steps: treatment preparation, which consists of patient simulation and treatment planning, and treatment execution, which delivers radiation

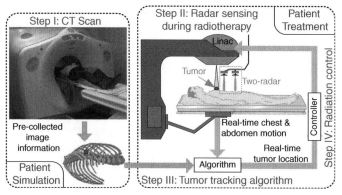

Figure 5.15 Motion-adaptive radiotherapy using continuous wave radar to track the location of tumor in real time. From Gu et al. (2012a).

dose to the patient. At the patient simulation stage, the patient and tumor geometrical information is collected through computed tomography (CT) scan and then a 3D patient model is built for the target tumor and organs at risk (Jiang, 2006a; Jiang, 2006b). The patient's breathing pattern is also examined at this stage. Those patients who cannot exhibit stable breathing patterns would be excluded from the motion-adaptive radiotherapy. Treatment planning is a virtual process that designs the patient treatment using the patient model built at the simulation stage. During the radiotherapy execution stage, a medical linear accelerator (LINAC) would work together with two radar sensors that dynamically monitor the chest wall and the abdomen to provide the real-time motion information. The LINAC could also be integrated with a radar sensor having beam-scanning capability, shown as inset b, which makes it possible to use one radar sensor to simultaneously measure the breathing motions at multiple body locations. In the third step, advanced tumor tracking algorithm combines the chest wall and abdomen motion information together with the precollected patient model to extract the tumor locations in real time. Then a controller utilizes the extracted tumor location information to control the LINAC to either perform gated radiotherapy or steer the radiation beam to track the tumor.

As a demonstration, a radar sensor was designed with a DC-coupled adaptive tuning architecture that includes RF coarse tuning and baseband fine-tuning. The RF tuning was implemented using a path of an attenuator and a phase shifter at the RF front-end of the radar sensor (Mostafanezhad and Boric-Lubecke, 2011; Pan et al., 2011). It adds a portion of the transmitter signal to the receiver signal to cancel most of the DC offset. To further calibrate the remaining DC offset, the baseband fine-tuning architecture was used to adaptively adjust the amplifier bias to the desired level that allows both high gain amplification and maximum dynamic range at the baseband stage. With the above-mentioned DC-tuning architectures, the radar sensor is able to precisely measure low frequency respiration motions with stationary moment.

The radar sensor was tested in a lab environment to demonstrate its ability of accurate displacement measurement that preserves DC information of stationary moment. Moreover, the radar sensor was integrated and tested with LINAC to validate its clinical use, as shown in Fig. 5.16. The radar measured respiration is shown in Fig. 5.17. On

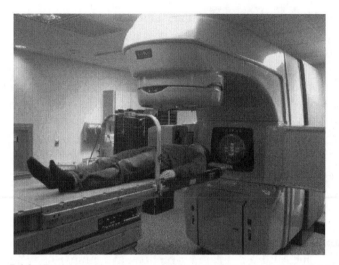

Figure 5.16 Radar respiratory monitoring setup with a human subject simulating patient breathing.

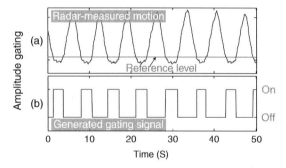

Figure 5.17 Radar measured respiration signal (a) and generated gating signal (b). The shaded area indicates the coach reference area, and the red line is the reference for gating.

the basis of the radar output, the subject was coached to dynamically adjust his breathing to put the EOE position within the shaded area, so as to generate reproducible respiration signals, from which, gating signals would be easy to be obtained. The accurate measurement of the DC-coupled radar sensor also allows the coaching reference area to be chosen near the position of end of inspiration (EOI), depending on specific clinical situations. The red line in Fig. 5.17a shows the reference for amplitude gating. The respiration signal triggers the radiation on once its amplitude falls below the reference line. Figure 5.17b shows the gating signals with a duty cycle of 40.5%. This experiment along with two experiments on Respiratory Gating Platform (Standard Imaging, Middleton, WI) (Gu et al., 2012a) has demonstrated the feasibility of using CW radar to precisely measure respiration pattern in motion-adaptive radiotherapy.

It should be noted that, in order to achieve accurate radiation beam targeting using respiration measurement, a correlation model needs to be built and validated between the internal tumor target and external respiration signal. This can be decoupled and is generally treated as a separate issue from respiration measurement. In fact, various approaches have been explored to infer the internal target position from external measurement. Interested authors are encouraged to refer to Yan et al. (2006) and Berbeco et al. (2006).

5.1.8 Structural Health Monitoring

Wireless smart sensor technology has recently drawn many interests in infrastructure monitoring and maintenance by providing pertinent information regarding the condition of a structure at a lower cost and higher density than traditional monitoring approaches. Many civil structures, especially long-span bridges, have low fundamental response frequencies that are challenging to accurately measure with sensors that are suitable for integration with low cost, low profile, and power-constrained wireless sensor networks. Existing displacement sensing technologies, such as LVDT, accelerometer, and global positioning system (GPS), are either not practical for wireless sensor implementations (LVDT), does not provide the necessary accuracy (accelerometer and GPS), or are simply too cost prohibitive (GPS) for dense sensor deployments. By providing accurate, low cost, noncontact sensing of vibration and displacement, radar motion sensor provides an opportunity to enhance low frequency vibration-based structural health monitoring and the measurement of static structural deflections.

Vibration-based structural health monitoring requires the ability to reconstruct accurate dynamic models of the structure being monitored. These models may be created using vibration measurements from sensors distributed at discrete points throughout the structure. Using bridge monitoring as an example, Fig. 5.18 compares three

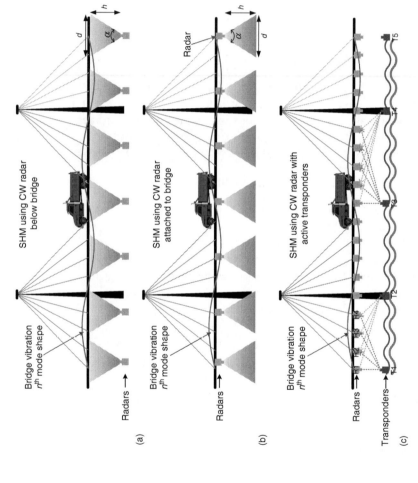

Figure 5.18 (a, b, c) Three configurations of radar structural vibration and displacement sensor. From Rice et al. (2011) and Gu et al. (2012b).

approaches of vibration measurement. If a radar is placed below the bridge as shown in Fig. 5.18a, generating a wave to be reflected from the moving surface on the underside of the bridge, the signal from the radar is dispersed and reflected off a region of the underside of the bridge with diameter, d, rather than a single point on the structure. The diameter of the measured region is given by $d = 2h \tan(\alpha)$, where h is the distance between the sensor and the target surface and α is the angle of dispersion. Therefore, measured motion of the bridge will be an average of the motion of the region, and not an accurate measurement of a discrete point on the bridge. This is particularly problematic when trying to measure higher modes of vibration, where the motion in a region of the bridge may have components that cancel one another out. To avoid the introduction of error caused by finite antenna directivity in bridge vibration measurement, the radars will be placed on the underside of the structure and transmit their signal to the surface below the structure, as shown in Fig. 5.18b. Assuming that the ground or water surface is still, the resulting measurements will accurately capture the motion at the locations of the sensors on the structure. If the ground or water surface below the bridge has motion, it will cause error to measurement in backscattering mode. In that case, active transponder discussed in Chapter 3 can be placed below the bridge to eliminate clutter noise, as shown in Fig. 5.18c.

Laboratory-scale tests have been performed on a two-storey building model and a truss bridge model. In Rice et al. (2011), a 1-m tall, two-storey building model (shown in Fig. 5.19) was instrumented with a radar sensor and an accelerometer (PCB 333B50) at each storey. An impulse load was applied to the first floor of the building, which was allowed to vibrate freely. The expected motion of the building is harmonic with two fundamental frequencies of vibration present. Operating in the arctangent demodulation mode to fully recover the motion pattern, the radar transmit signal toward a vertical screen positioned 2.8 m from the original position of the building.

Representative displacement data from the radar sensors is shown in Fig. 5.20, with the amplitudes of vibration approximately 3 and 2 mm for the first and second floor, respectively. To compare the data from the radar sensors and the accelerometers in the time domain, the displacements from the radar sensors were differentiated twice and filtered with a low pass filter to produce an acceleration

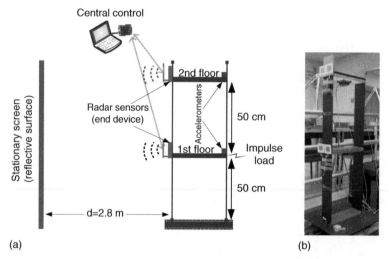

(a)   (b)

Figure 5.19 Two-story building test configuration (a) and picture of building model with radar sensors installed on each floor (b). From Rice et al. (2011).

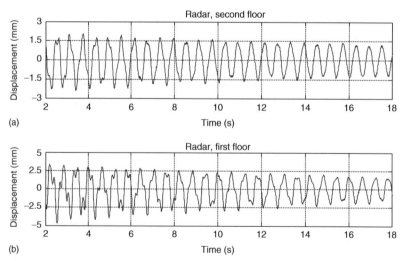

Figure 5.20 Displacement measured by radar installed on each floor. From Rice et al. (2011).

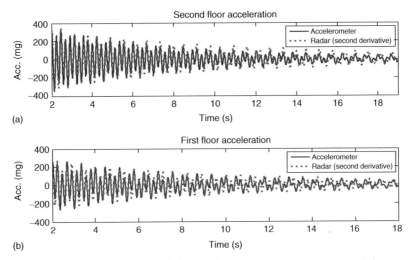

Figure 5.21 Time histories of the accelerometer measurements and the second derivative of the radar measurements. From Rice et al. (2011).

record, which is shown in Fig. 5.21. The spectrum and resulting operational deflection shapes (ODSs) are shown in Fig. 5.22. The ODSs were calculated using the peak picking method (Silva, 2007).

The signals match well in the time domain. In the frequency responses, the peaks representing the fundamental frequencies of vibration match perfectly; however, some discrepancy is seen in the amplitudes of the peaks. This error may be the result of measurement error in the arctangent demodulated interferometry mode. Additional error may be introduced through the differentiation of the radar signal. The difference in the spectrum amplitudes is further evident in the comparison of the ODSs: while the second ODSs match very well, the first ODSs are slightly different, possibly because there is less energy in the first mode, resulting in a lower signal-to-noise ratio at that frequency. In addition, the spectrum of the radar data shows additional peaks at harmonics of the first natural frequency. These harmonics are introduced by the modulation/demodulation algorithm when the DC offset is not perfectly extracted. While the amplitude of the spurious harmonics is small relative to the true fundamental frequencies, they are still a concern for accurate modal analysis. Further signal processing refinement, such as bandpass filtering, is necessary to eliminate the unwanted peaks.

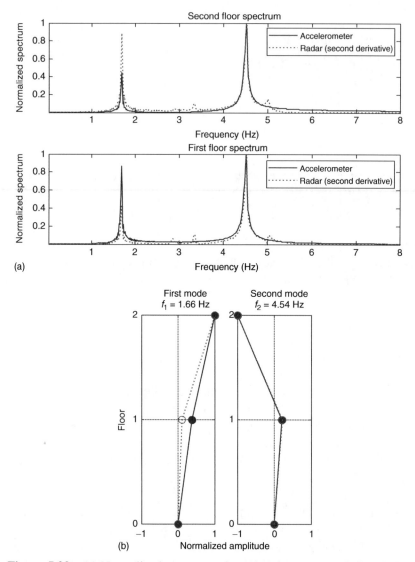

Figure 5.22 (a) Normalized spectrum of acceleration measured directly by the accelerometers and from the second derivative of the radar measurements. (b) Normalized operational deflection shapes from accelerometer and radar measurements. From Rice et al. (2011).

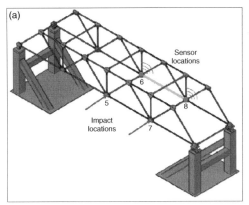

Figure 5.23 (a) Test configuration on the truss bridge model. (b) Picture of a radar sensor on node 8. From Rice et al. (2012).

To demonstrate the feasibility of using wireless radar sensors for bridge monitoring, a 3-m long, three-dimensional truss bridge was instrumented with radar sensors and accelerometers collocated at nodes 6 and 8, and an LVDT at node 6, as shown in Fig. 5.23. The bridge was excited by horizontal hammer impacts at either node 5 or 7. A wall approximately 400 mm from the plane of the bridge on which the sensors were located acted as the target for the radar sensors.

Representative displacement data from a lateral impact force consecutively imposed at node 5 is shown in Fig. 5.24. The amplitudes of vibration at node 6 reach approximately 2 mm while the amplitudes at node 8 are less than 1 mm (smaller amplitudes than those measured in the shake table tests). To compare the data from the radar sensors and the accelerometers in the time domain, the accelerations were filtered with a high pass filter and integrated twice to produce displacement records. In addition, an LVDT was mounted to measure the lateral displacement at node 6.

All three measurement signals match well in the time domain. An example of the difference between the radar data and the integrated accelerometer data is shown in Fig. 5.25. The average root mean square (RMS) difference between the integrated accelerometer data and the radar sensors is less than 0.2 mm. The power spectral densities are shown in Fig. 5.26. The peak values for each sensor type,

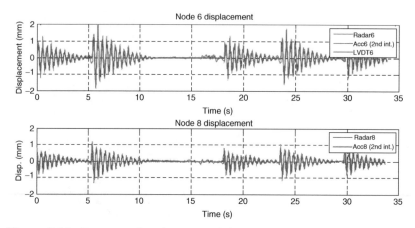

Figure 5.24 Response data from truss bridge impact tests. From Rice et al. (2012).

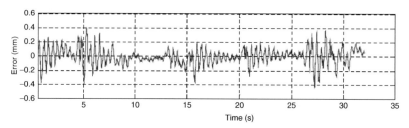

Figure 5.25 Difference plot between radar measurement and double-integrated acceleration measurement. From Rice et al. (2012).

representing the lateral and torsional natural frequencies of the bridge, match well. The amplitudes of LVDT and radar sensor match very well up to 5 Hz, while the integrated accelerometer amplitude and the radar data match well from approximately 2 Hz onward. One reason for the poor performance of the accelerometer in the low frequency range is that ICP accelerometers (PDB 333B50) with a frequency range of 0.5–3000 Hz were used. Additionally, the LVDT has a range of measurement from DC to 5 Hz for displacements less than 5 mm.

The results in the bridge monitoring study verified the sensor performance on a laboratory-scale bridge structure, demonstrating the validity of the radar sensor measurements with submillimeter accuracy (Rice et al., 2012). Figure 5.27 compares the frequency response

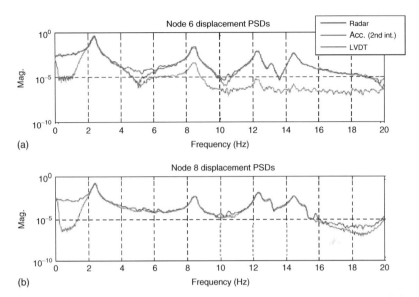

Figure 5.26 Power spectral densities of bridge response measured by accelerometers (integrated twice) and radar sensors at nodes 6 and 8 and a linear displacement sensor at node 6. From Rice et al. (2012).

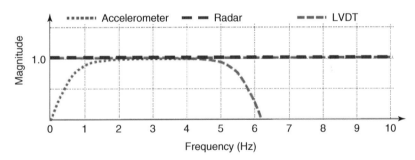

Figure 5.27 Comparison of frequency responses of LVDT, accelerometer, and radar used in the bridge monitoring study. From Rice et al. (2012).

capabilities of each of the sensor types used in this study. While the other sensors have bandwidth limitations, the radar sensor accurately captures motion from DC up to 20 Hz and beyond.

Furthermore, technologies discussed in this book can construct multifunction interferometric radar sensor that can easily form a smart sensor network with the help of the ZigBee mesh networking function integrated in the radar motion and displacement sensor. Gu et al. demonstrated a radar sensor that monitors a structure's displacement with an accuracy of submillimeter (Gu et al., 2012b). Many large structures, such as the Golden Gate Bridge with a length of 2.7 km and a height of 227.4 m, require a dense array of sensors to cover the large physical area. Taking a long-span bridge as an example, a smart wireless sensor network (WSSN) using multifunction radar motion sensors is shown in Fig. 5.28. The cosine-like line shows the low frequency fundamental vibration of the long-span bridge. The radar sensors are mounted on the underside of the bridge so that the sensor vibrates along with the bridge's vibration. When mounted on discrete points on the bridge, the radar sensors can accurately capture the motion at their specific locations. Experimental results have shown that the smart sensor network using the multifunction radar sensors serves as a good alternative for monitoring structural health. Continuing works with practical installation and field tests are in progress.

5.2   DEVELOPMENT OF STANDARDS AND STATE OF ACCEPTANCE

Various applications of the noncontact microwave motion sensing technology correspond to different standards and regulations. Currently, there is no universal standard developed for microwave motion sensing radar. For healthcare applications such as sleep apnea and SIDS monitoring, the devices must obtain FDA (U.S. Food and Drug Administration) approval for use in U.S. and equivalent approvals for use in other countries before they are available to public. For industrial applications, the devices have to be tested for EM compatibility with other instruments. The carrier frequency of the radar sensor has to be properly chosen to meet all the radio spectrum requirements.

A popular choice for the frequency of many microwave motion sensor prototype is the ISM bands, which stands for the "industrial, scientific, and medical radio bands." They are radio bands, or portions of the radio spectrum, reserved internationally for the use of RF energy for ISM purposes other than licensed communications. Examples of applications in these bands include medical diathermy machines, RF process heating, and microwave ovens. The powerful emissions of

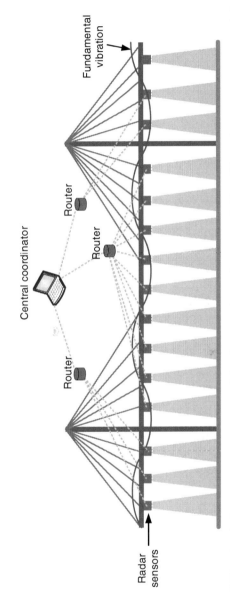

Figure 5.28 A wireless smart sensor network implemented on a long-span bridge. The radar sensors are mounted on the underside of the bridge (Gu et al., 2012b).

195

these devices can create EM interference and disrupt radio communication using the same frequency, so these devices were limited to certain bands of frequencies. However, radar sensors operating in these bands must tolerate any interference generated by ISM band equipment and devices, such as Wireless Fidelity (Wi-Fi), ZigBee, and Bluetooth handsets.

A general concern for civilian microwave sensing applications is the safety of EM radiation. It should be noted that the transmitted RF power of most of the systems and applications discussed in this book is lower than 1 W, which is lower than the peak transmit power of a conventional Global System for Mobile Communication (GSM) phone. It is safe in terms of long-term human exposure to RF signals according to the ANSI C95.1-1991 Standard as well as other standards/regulations/reports.

5.3 FUTURE DEVELOPMENT TRENDS

From remotely monitoring an infant's breathing to detecting the spin speed of a fast-rotating MEMS disk, the wide range of ways by which remote radar motion sensing technology can be used to benefit people is astounding. Although the nonintrusive approach, high accuracy, and attractive features indicate that it will eventually become a part of daily life for many people, the technology is still in the developmental phase. Several development trends are expected from the following aspects.

While board level systems built from off-the-shelf components provide a fast prototype cycle, miniature-sized high sensitivity radar-on-chip will be continuously developed for specific applications because of the low cost and low power feature. CMOS process will be a good candidate for the radar-on-chip system integration in most of the low cost daily applications. BiCMOS and III–V compound semiconductor process may be adopted for high end military or astronautic applications such as through-the-wall enemy detection and structural health monitoring.

The carrier frequency of radar motion system will be pushed to millimeter wave or even Terahertz range for higher sensitivity with a shorter wavelength. Preliminary works have already been demonstrated in millimeter-wave range. For example, Laskin et al. reviewed recent development conducted at the University of Toronto on the development of CMOS transceivers aimed at operation in the 90–170 GHz range (Las et al., 2009). Transistor and top-level layout issues, low voltage circuit topologies, and design flow were addressed for millimeter-wave circuit design in the 65-nm node and beyond. Transceivers with both quadrature architecture and double-sideband architecture were fabricated in a 65-nm general purpose and low power (GPLP) CMOS technology. The double-sideband transceiver was intended for the remote monitoring of respiratory functions. A Doppler shift of 30 Hz, which was produced by a slow-moving (4.8 cm/s) target located at a distance of 1 m, was measured with a transmitter output power of approximately 2 dBm and a phase noise of −90 dBc/Hz at 1 MHz offset. The range correlation effect was demonstrated by measuring the phase noise of the received baseband signal at 10-Hz offset. Measurement result verified that the $1/f$ noise has been canceled and it does not pose a problem in short-range motion sensing applications. With the fast development of Terahertz technology, the radar motion sensor is attractive for microscale sensor in monitoring deflection, linear motion, actuation, and rotational movement of miniaturized structures such as MEMS devices.

As the carrier frequency increases, microwave/millimeter-wave interferometry is drawing both academic and industrial interests. For example, it was recently proposed in Doronin et al. (2012) that a two-probe measurement could be implemented for precise displacement measurement, realizing a theoretical worst-case accuracy of about 4.4% of the operating wavelength. In Mikhelson et al. (2012b), a 94-GHz millimeter-wave interferometer was used to obtain the relative displacement of an object. The physiological motions of the chest due to cardiac activity were measured using the interferometer system. After processing the data using a wavelet multiresolution decomposition, a signal with peaks at heartbeat temporal locations was obtained. It was found out that, in order for these heartbeat temporal locations to be accurate, the noise in the reflected signal must be well controlled. A statistical algorithm was created to compensate for unconfident temporal locations as computed by the wavelet transform. By analyzing the statistics of the peak locations, missing heartbeat temporal locations were filled in and superfluous ones were eliminated. With the above methods adopted, the system is able to find the HR of ambulatory subjects without any physical contact.

In Diebold et al. (2012), an MMIC CW radar system operating at 96 GHz was reported. The MMIC components were custom-built in 100 nm metamorphic high electron mobility transistors (mHEMTs) technology. The radar system employed a multiply-by-twelve frequency multiplier MMIC and a receiver MMIC both packaged in split-block modules. The radar was able to detect respiration and heartbeat frequency of a human target sitting at 1 m in front of the radar. Both time- and frequency-domain analyses were carried out. The advantages and drawbacks of each approach were compared and summarized.

An emerging application of microwave/millimeter-wave radar motion sensor takes advantage of the resonance modes of materials, which are intrinsic characteristics of objects when excited at certain frequencies. Probing the resonance signatures can reveal useful information about material composition, geometry, presence of defects, and other characteristics of the object under test (Liao et al., 2012), leading to promising usage in nondestructive evaluation of materials. On the basis of the technologies discussed in this book, vibration spectra can be remotely measured with high degree of sensitivity using a millimeter-wave radar. This results in a novel nondestructive evaluation method that can work in a noncontact manner as an alternative or complementary approach to conventional methods based on acoustic/ultrasonic and optical techniques. Some civilian and military/security examples include detection of defects and degradation for diagnostics and prognostics of materials components and rapid standoff inspection of shielded/sealed containers for contraband.

In Liao et al. (2012), the performance of a compact millimeter-wave vibrometer developed at Argonne National Lab was evaluated. The 94 GHz $I - Q$ Doppler sensor successfully monitored the mechanical vibration signature of the object under interrogation that was induced by CW excitation. For proof-of-principle demonstrations, the test objects were mechanically excited by an electronically controlled shaker using sinusoidal waves at various frequencies ranging from DC to 200 Hz. The experimental setup is shown in Fig. 5.29. The radar signal generated by a W-band solid-state Gunn oscillator is transmitted through a corrugated horn antenna (2.39 mm in diameter) and is focused by a 6-in.-diameter dielectric lens on the target.

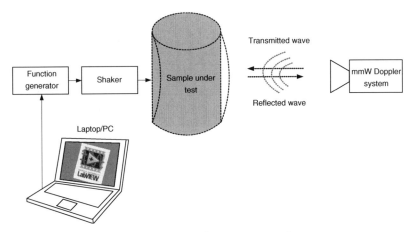

Figure 5.29 Experimental setup for resonant signature measurement. Adapted from Liao et al. (2012).

It should be noted that millimeter-wave Doppler vibrometer has the advantages of high sensitivity and high spatial resolution (with a reasonable aperture size) because of the short wavelength. Compared with laser Doppler technique, the microwave/millimeter-wave approach has the unique advantage of penetration through many optically opaque dielectric materials, low atmospheric attenuation, and low sensitivity to surface condition of the object. Ongoing investigations in Argonne National Laboratory are associated with incorporation of an air-coupled excitation source such as high power speaker or a laser, which may enable the method to be implemented as a standoff non-destructive evaluation tool (Liao et al., 2012).

Although the microwave noncontact motion sensing approach is ideal for various biomedical applications, many efforts are needed on clinical study and comparison with existing contact-based methods such as electrocardiogram. Preliminary works have been done on various subjects and in different environment. For example, Yan et al. performed experiments with infant simulators to demonstrate the efficacy of noncontact vital sign detection radar in children and baby healthcare in the Neonatal Intensive Care Unit (NICU) at Shands Hospital at the University of Florida (Yan et al., 2009). A 5.8 GHz Doppler radar was placed below an infant crib where X-rays can be taken through a nonmetal structure. The radio wave can also penetrate through and reach the infant simulator. The experiment was conducted through 18 continuous events. In each event, RR, HR, and tidal volume (TV) were controlled and changed by software. The experimental results proved that the radar sensor could achieve accurate detection results when the infant simulator was in a baseline or normal physiologic state. Although the results show deviations to the programmed values when detecting the infant under abnormal conditions, they can reflect the change of vital signs under abnormal conditions and provide approximate data that are useful for initial diagnose of specific syndromes. Detailed analyses of the typical cases can be found in Yan et al. (2009).

The efficacy of Doppler radar as a substitute for conventional respiratory monitors has also been demonstrated in human population studies. Testing conducted using a spirometer simultaneously with Doppler radar has indicated that displacement-based TV measurements are possible with the subject in a known position. Massagram et al. carried out a short-term measurement on 10 awake supine subjects (Massagram et al., 2009), where TV was obtained using a 2.4 GHz quadrature Doppler radar system, and compared with a spirometer. The study indicated that it was possible to establish a correlation parameter between chest displacement and TV for each subject. Using a Bland–Altman plot for spirometer versus radar TV measurements, the difference between the two techniques was analyzed, revealing the occurrence that the two methods can be considered an equivalent substitute for the other. In another study (Droitcour et al., 2009), clinical results were obtained validating the accuracy of RR obtained for hospitalized patients using a noncontact, low power 2.4 GHz Doppler radar system. Twenty-four patients were measured, and the RR accuracy was benchmarked against the RR obtained using a Welch Allyn Propaq Encore model 242, the Embla Embletta chest-belt system, and by counting chest excursions. The 95% limits of agreement between the Doppler radar and reference measurements fell within ±5 breaths/min.

Techniques for the precise assessment of key parameters relating to cardiopulmonary activity will continue to be investigated. A pioneering study in Kiriazi et al. (2012) used dual-band CW radar system operating at 2.4 and 5.8 GHz. Quantitative analyzes of the return signal, in terms of intensity and phase modulation magnitude were performed. On the basis of the study, the first parameter, defined as the cardiopulmonary effective radar cross section (ERCS), was introduced to measure the radar cross section of the portion of the torso surface that moves due to respiration and heartbeat activity. The second parameter corresponds to the maximum displacement of the torso surface in the direction of incidence. Both the cardiopulmonary ERCS and displacement were determined using the dual-band radar system. A calibration procedure using spherical targets was provided and displacement measurement was validated for the dual-frequency system. Preliminary testing results consistently showed an ERCS that is larger for the back of the torso and smaller for the side compared to the front, while the respiration depth is smaller in the prone position than in supine. It was found that determination of body orientation is essential for displacement-based measurements such as TV. With ERCS tracking, a radar system can automatically determine orientations and therefore apply the correct calibration relating displacement to TV.

On the basis of further radar signal processing, microwave Doppler radar can reveal many characteristics of human and animal motion, including gaits and biometrics information. This information can be used to detect humans and distinguish them from natural fauna, which is an important issue in border security applications. In particular, it is valuable to detect and classify people who are walking in remote locations over extended periods at low operating and maintaining costs. For example, Tahmoush et al. used simulation and measurement to demonstrate a reasonable model for studying signatures using micro-Doppler radar (Tahmoush and Silvious, 2009). Kim and Ling used micro-Doppler signatures to classify human activities (Kim and Ling, 2009). Micro-Doppler signals refer to Doppler scattering returns produced by the motions of the target other than gross translation. Some small micro-Doppler signatures returns from parts of a human body are periodic and therefore can illustrate gaits characteristics using analytical techniques. Micro-Doppler sensing gives rise to many detailed radar image features in addition to the information associated with the bulk target motions. Researchers have been investigating modulations of the radar return from arms, legs, and body sway. Interested readers can refer to Chen and Ling (2002), Chen and Lipps (2000b), and Chen et al. (2006) for state of the art and some tutorials on micro-Doppler phenomena.

5.4   MICROWAVE INDUSTRY OUTLOOK

The abundant applications indicate great potential for microwave industry to have a prospective market on the motion sensing technologies discussed in this book. The impacts of these technologies extend to fields including healthcare, home security, national security, energy efficiency, smart house, industrial control, and civil engineering. Multiple startup companies have been established in recent years over the world. In the meantime, plenty of patents have been published as application or awarded in the past decades on technology details such as wavelength division sensing, vibration monitor, random body movement noise cancelation, and respiratory monitoring. It is expected that products of this industry will be used in commercial, safety, rescue, and military purposes.

In the future, the competition in this field will be ruled by patent protection of a product, government approval for the technology used, and product specifications such as accuracy, resolution, and power consumption. As some applications are directly used on humans, it is mandatory for a company to receive government approval to use the product. As an example, in order to be commercialized in the United States, a radar respiratory monitor has to receive clearance from the section 510(k) of the Food, Drug and Cosmetic Act, which requires device manufacturers who must register, to notify FDA of their intent to market a medical device at least 90 days in advance.

From remote monitoring the vital sign of a newborn baby to detection of enemy through an obstacle in battle field, tracking tumor during radiotherapy, finding buried earthquake victims, or monitor the condition of building structures, the wide range of ways radar noncontact motion sensing can be used is astounding. Although, the technology is still in the development phase, the nonintrusive approach, high accuracy, and attractive features indicate that it will eventually create a great market for the microwave industry.

REFERENCES

Adler R. A study of locking phenomena in oscillators. Proc IRE 1946;34(6): 351–357.

Asmar R. *Arterial Stiffness and Pulse Wave Velocity: Clinical Application*. Paris, France: Elsevier; 1999.

Bakhtiari S, Elmer TW, Cox NM, Gopalsami N, Raptis AC, Liao S, Mikhelson I, et al. Compact millimeter-wave sensor for remote monitoring of vital signs. IEEE Trans Instrum Meas 2012a;61:830–841.

Balanis CA. *Antenna Theory: Analysis and Design*. 3rd ed. Wiley-Interscience, Hobokan, NJ, USA; 2005.

Benlarbi A, Van De Velde JC, Matton D, Leroy Y. Position, velocity profile measurement of a moving body by microwave interferometry. IEEE Trans Instrum Meas 1990;39:632–636.

Berbeco RI, Nishioka S, Shirato H, Jiang SB. Residual motion of lung tumors in end-of-inhale respiratory gated radiotherapy based on external surrogates. Med Phys 2006;33(11).

Boric-Lubecke O, Awater G, Lubecke VM. Wireless LAN PC card sensing of vital signs, 2003 IEEE AP-S Topical Conference on Wireless Communications Technology; Honolulu (HI); 2003 Oct.

Microwave Noncontact Motion Sensing and Analysis, First Edition.
Changzhi Li and Jenshan Lin.
© 2014 John Wiley & Sons, Inc. Published 2014 by John Wiley & Sons, Inc.

Boric-Lubecke O, Lubecke V, Host-Madsen A, Samardzija D, Cheung K. Doppler radar sensing of multiple subjects in single and multiple antenna systems, 7th International Conference on Telecommunications in Modern Satellite, Cable and Broadcasting Services, vol. 1; 2005 Sep. p. 7–11.

Boric-Lubecke O, Lin J, Park B-K, Li C, Massagram W, Lubecke VM, Host-Madsen A. Battlefield triage life signs detection techniques, Proceedings of the SPIE Defense and Security Symposium, Vol. 6947—Radar Sensor Technology XII, No. 69470J, 10 pages, 2008 Apr.

Chen KM, Misra D, Wang H, Chuang HR, Postow E. An X-band microwave life-detection system. IEEE Trans Biomed Eng 1986;33:697–702.

Chen KM, Kallis J, Huang Y, Sheu JT, Norman A, Lai CS, Halac A. EM wave life-detection system for post-earthquake rescue operation, presented at the 1994 URSI Radio Science Meeting; Seattle (WA); 1994 Jun 19–24.

Chen KM, Huang Y, Norman A, Yerramille Y. EM wave life-detection system for post-earthquake rescue operation-field test and modifications, Proceedings of the 1996 IEEE/APS-URSI International Symposium; Baltimore (MD); 1996 Jul 21–26.

Chen VC, Ling H. Joint time-frequency analysis for radar signal and image processing. IEEE Signal Proc Mag 1999;16(2):81–93.

Chen KM, Huang Y, Zhang J, Norman A. Microwave life-detection systems for searching human subjects under earthquake rubble and behind barrier. IEEE Trans Biomed Eng 2000a;27:105–114.

Chen VC, Lipps R. Time frequency signatures of micro-Doppler phenomenon for feature extraction, Proceedings of SPIE, vol. 4056, Wavelet Applications VII; 2000b Apr. p. 220–226.

Chen VC, Ling H. *Time-Frequency Transforms for RadarImaging and Signal Analysis*. Norwood (MA): Artech House; 2002.

Chen VC, Li F, Ho S-S, Wechsler H. Micro-Doppler effect in radar: phenomenon, model, and simulation study. IEEE Trans Aeros Electron Syst 2006;42:2–21.

Chernyak VS, Immoreev I. A brief history of radar. IEEE Aeros Electronic Syst Mag 2009;24(9):B1–B32.

Choi SH, Park JK, Kim SK, Park JY. A new ultra-wideband antenna for UWB applications. Microw Opt Technol Lett 2004;40(5):399–401.

Choi J, Chung K, Roh Y. Parametric analysis of a band-rejection antenna for UWB application. Microw Opt Technol Lett 2005;47(3):287–290.

Chuang HR, Chen YF, Chen KM. Automatic clutter-canceller for microwave life-detection system. IEEE Trans Instrum. Meas 1991;40(4):747–750.

Curto S, John M, Ammann M. Groundplane dependent performance of printed antenna for MB-OFDM-UWB, Vehicular Technology Conference; Baltimore (MD); 2007 Apr. p. 352–356.

Diebold S, Ayhan S, Scherr S, Massler H, Tessmann A, Leuther A, Ambacher O, Zwick T, Kallfass I. A W-band MMIC radar system for remote detection of vital signs. J Infrared Millimeter Terahertz Waves 2012.

Doronin AV, Gorev NB, Kodzhespirova IF, Privalov EN. Displacement measurement using a two-probe implementation of microwave interferometry. Prog Electromag Res C 2012;32:245–258.

Dragoman D, Dragoman M. Terahertz fields and applications. Prog Quantum Electron 2004;28(1):1–66.

Droitcour AD, Lubecke V, Lin J, Boric-Lubecke O. A microwave radio for Doppler radar sensing of vital signs. IEEE MTT-S Int Microw Symp Dig; Phoenix (AZ); 2001:175–178.

Droitcour AD, Boric-Lubecke O, Lubecke V, Lin J. 0.25 µm CMOS and BiCMOS single chip direct conversion doppler radars for remote sensing of vital signs, IEEE International Solid State Circuits Conference Digest of Technical Papers; San Francisco (CA); 2002 Feb. p. 348–349.

Droitcour AD, Boric-Lubecke O, Lubecke V, Lin J, Kovacs G. Range correlation effect on ISM Band I/Q CMOS radar for non-contact vital signs sensing. IEEE MTT-S Int Microw Symp Dig 2003; Philadelphia (PA); 3:1945–1948.

Droitcour AD, Boric-Lubecke O, Lubecke V, Lin J, Kovac G. Range correlation and I/Q performance benefits in single-chip silicon Doppler radars for noncontact cardiopulmonary monitoring. IEEE Trans Microw Theory Tech 2004a;52:838–848.

Droitcour AD, Boric-Lubecke O, Lubecke V, Lin J, Kovacs G. Physiological motion sensing with modified silicon base station chips. IEICE Trans Electron 2004b;E87-C(9):1524–1531.

Droitcour AD, Seto TB, Byung-KwonP, Yamada S, VergaraA, El Hourani C, Shing T, et al. Non-contact respiratory rate measurement validation for hospitalized patients, in Engineering in Medicine and Biology Society, 2009. EMBC 2009. Annual International Conference of the IEEE; Minneapolis (MN); 2009, p. 4812–4815.

Fletcher R, Han J. Low-cost differential front-end for Doppler radar vital sign monitoring. IEEE MTT-S Int Microw Symp Dig 2009:1325–1328.

Gerald CF, Wheatley PO. *Applied Numerical Analysis*. 3rd ed. Reading (MA): Addison-Wesley; 1984.

Goupillaud P, Grossman A, Morlet J. Cycle-octave and related transforms in seismic signal analysis. Geoexploration 1984;23:85–102.

Greenberg MC, Virga L, Hammond CL. Performance characteristics of the dual exponentially tapered slot antenna (DETSA) for wireless communications applications. IEEE Trans Veh Technol 2003;52(2):305–312.

Gu C, Li C, Huangfu J, Lin J, Ran L. Instrument-based Non-contact Doppler Radar Vital Sign Detection System Using Heterodyne Digital Quadrature Demodulation Architecture. IEEE Transactions on Instrumentation and Measurement 2010;59(6):1580–1588.

Gu C, Li R, Zhang H, Fung AYC, Torres C, Jiang SB, Li C. Accurate respiration measurement using dc-coupled continuous-wave radar sensor for motion-adaptive cancer radiotherapy. IEEE Trans Biomed Eng 2012a;59(11):3117–3123.

Gu C, Rice J., et al. A wireless smart sensor network based on multi-function interferometric radar sensors for structural health monitoring, IEEE Radio and Wireless Week; Santa Clara (CA); 2012b Jan 15–18.

Gu C, Li C. DC coupled CW radar sensor using fine-tuning adaptive feedback loop. IET Electron Lett 2012c;48(6):344–345.

Gu C, Rice J, Li C. Interferometric radar sensor with active transponders for signal boosting and clutter rejection in structural health monitoring, IEEE International Microwave Symposium (IMS); Montreal, Canada; 2012d Jun 17–22.

Guarnieri M. The early history of radar [Historical]. IEEE Ind Electron Mag 2010;4(3):36–42.

Huang J. Planar microstrip Yagi array antenna. 1989 IEEE Antennas and Propag Soc Int Symp Dig 1989;2:894–897.

Huang J, Densmore AC. Microstrip Yagi array antenna for mobile satellite vehicle application. IEEE Trans Antennas Propag 1991;39(7):1024–1030.

Huey-Ru C, Hsin-Chih K, Fu-Ling L, Tzuen-Hsi H, Chi-Shin K, Ya-Wen O. 60-GHz millimeter-wave life detection system (MLDS) for noncontact human vital-signal monitoring. Sensors J IEEE 2012;12:602–609.

Huntoon RD, Weiss A. Synchronization of oscillators. Proc IRE 1947;35(12): 1415–1423.

Hynes M, Wang H, McCarrick E, Kilmartin L. Accurate monitoring of human physical activity levels for medical diagnosis and monitoring using off-the-shelf cellular handsets. Pers Ubiquitous Comput 2011;15:667–678.

Immoreev I, Samkov S, Tao T-H. Short-distance ultra-wideband radars. IEEE Aeros Electron Syst Mag 2005;20(6):9–14.

Immoreev I, Tao T-H. UWB radar for patient monitoring. IEEE Aeros Electron Syst Mag 2008;23(11):11–18.

James RJ. A history of radar. IEE Rev 1989;35(9):343–349.

Jiang SB. Radiotherapy of mobile tumors. Semin Radiat Oncol 2006a;16(4): 239–24.

Jiang SB. Technical aspects of image-guided respiration gated radiation therapy. Med Dosim 2006b;31(2):141–151.

Kao TYJ, Chen AYK, Yan Y, Tze-Min S Jenshan L. A flip-chip-packaged and fully integrated 60 GHz CMOS micro-radar sensor for heartbeat and mechanical vibration detections, IEEE Radio Frequency Integrated Circuits Symposium (RFIC); 2012. p. 443–446.

Kim H, Ling H. Wavelet analysis of radar echo from finite-size targets. IEEE Trans Antennas Propag 1993;41(2):200–207.

Kim S, Nguyen C. A displacement measurement technique using millimeter-wave interferometry. IEEE Trans Microw Theory Tech 2003;51: 1724–1728.

Kim S, Nguyen C. On the development of a multifunction millimeter-wave sensor for displacement sensing and low-velocity measurement. IEEE Trans Microw Theory Tech 2004;52:2503–2512.

Kim Y and Ling H. Human activity classification based on microdoppler signatures using a support vector machine. IEEE trans. Geosci. Remote Sensing, 2009;47:1328–1337.

Kiriazi JE, Boric-Lubecke O, Lubecke VM. Dual-Frequency technique for assessment of cardiopulmonary effective RCS and displacement. Sensors J IEEE 2012;12:574–582.

Kumar G, Gupta KC. Nonradiating edges and four edges gap-coupled multiple resonator broad-band microstrip antennas. IEEE Trans Antennas Propag 1985;33(2):173–178.

Kurokawa K. Injection locking of microwave solid-state oscillators. Proc IEEE 1973;61(10):1386–1410.

Lai SHY. Engine system diagnosis using vibration data. Comp Ind Eng 1993;25(1–4):135–138.

Las E, Kha M, et al. Nanoscale CMOS transceiver design in the 90-179-GHz range. IEEE Trans Microw Theory Tech 2009;57(12):3477–3490.

Laurent F, Latrabe V, Vergier B, Montaudon M, Vernejoux J, Dubrez J. CT-guided transthoracic needle biopsy of pulmonary nodules smaller than 20 mm results with an automated 20-gauge coaxial cutting needle. Clin Radiol 2000;55:281–287.

Li J, Stoica P. Efficient mixed-spectrum estimation with applications to target feature extraction. IEEE Trans. Signal Proc 1996;44(2):281–295.

Li C, Xiao Y, Lin J. Experiment and spectral analysis of a low-power Ka-band heartbeat detector measuring from four sides of a human body. IEEE Trans Microw Theory and Tech 2006a;54(12):4464–4471.

Li C, Xiao Y, Lin J. Robust Overnight Monitoring of Human Vital Sign by a Non-contact Respiration and Heartbeat Detector. IEEE 28th Annual International conference of the IEEE Engineering in Medicine and Biology Society 2006b;2235–2238.

Li C, Lin J. Non-contact measurement of periodic movements by a 22-40 GHz radar sensor using nonlinear phase modulation, IEEE MTT-S International Microwave Symposium; Honolulu (HI); 2007a Jun. p. 579–582.

Li C, Lin J. Optimal carrier frequency of non-contact vital sign detectors, Proceedings of IEEE Radio and Wireless Symposium; Long Beach (CA); 2007b Jan 9–11. p. 281–284.

Li C, Lin J, Xiao Y. Design guidelines for radio frequency non-contact vital sign detection, IEEE 29th Annual International Conference of the IEEE Engineering in Medicine and Biology Society; Lyon, France; 2007c Aug 23–26; p. 1651–1654.

Li C, Lin J. Random Body Movement Cancellation in Doppler Radar Vital Signs Detection, IEEE Transactions on Microwave Theory and Techniques 2008a;56(12):3143–3152.

Li C, Xiao Y, Lin J. A 5-GHz double-sideband radar sensor chip in 0.18-μm CMOS for non-contact vital sign detection. IEEE Microw Wireless Compon. Lett. 2008b;18(7):494–496.

Li C, Cummings J, Lam J, Graves E, Wu W. Radar remote monitoring of vital signs. IEEE Microw Mag 2009a;10(1):47–56.

Li C, Lin J. Doppler radar non-contact measurement of rotational movement in both macro- and micro- scales, IEEE Asia-Pacific Microwave Conference (APMC); Singapore; 2009b Dec.

Li C, Yu X, Li D, Ran L, Lin J. Software configurable 5.8 GHz radar sensor receiver chip in 0.13 μm CMOS for non-contact vital sign detection. IEEE RFIC Symposium Digest of Papers; Boston (MA); 2009c:97–100.

Li C-J, Wang F-K, Horng T-S, Peng K-C. A novel RF sensing circuit using injection locking and frequency demodulation for cognitive radio applications. IEEE Trans Microw Theory Tech 2009d;57(12):3143–3152.

Li C, Ling J, Li J, Lin J. Accurate Doppler Non-contact Vital Sign Detection Using the RELAX Algorithm, IEEE Transactions on Instrumentation and Measurement 2010a;59(3):687–695.

Li C, Yu X, Lee C-M, Li D, Ran L, Lin J. High sensitivity software configurable 5.8 GHz radar sensor receiver chip in 0.13 μm CMOS for non-contact vital sign detection. IEEE Trans Microw Theory TechRFIC2009 Special Issue 2010b;58(5):1410–1419.

Li C, Gu C, Li R, Jiang SB. Radar motion sensing for accurate tumor tracking in radiation therapy, IEEE Wireless and Microwave Technology Conference; Clearwater Beach (FL); 2011 Apr.

Liao S, Bakhtiari S, Elmer T, Lawrence B, Koehl ER, Gopalsami N, Raptis A. Millimeter Wave Doppler Sensor for Nondestructive Evaluation of Materials, Proceedings of the ASNT 21st Annual Research Symposium; 2012.

Lin JC. Noninvasive microwave measurement of respiration. Proc IEEE 1975;63(10):1530.

Lin JC. Microwave sensing of physiological movement and volume change: a review. Bioelectromagnetics 1992;13:557–565.

Ling H, Moore J, Bouche D, Saavedra V. Wavelet analysis of radar echo from finite-size targets. IEEE Trans Antennas Propag 1993;41(8):1147–1150.

Liu Z, Li J. Implementation of the RELAX algorithm. IEEE Trans Aeros Electron Syst 1998;34(2):657–664.

Lubecke V, Boric-Lubecke O, Beck E. A compact low-cost add-on module for Doppler radar sensing of vital signs using a wireless communications terminal. IEEE MTT-S Int Microw Symp Dig 2002:1767–1770.

Mackey RC. Injection locking of klystron oscillators. IRE Trans Microw Theory Tech 1962;10(4):228–235.

Massagram W, Lubecke VM, Boric-Lubecke O. Microwave non-invasive sensing of respiratory tidal volume, in Engineering in Medicine and Biology Society, 2009. EMBC 2009. Annual International Conference of the IEEE; 2009. p. 4832–4835.

Meinel HH. Commercial applications of millimeter waves history, present status, and future trends. IEEE Trans Microw Theory Tech 1995;43(7):1639–1653.

Mikhelson IV, Bakhtiari S, Elmer TW II,, Sahakian AV. Remote sensing of patterns of cardiac activity on an ambulatory subject using millimeter-wave interferometry and statistical methods. Med Biol Eng Comput 2012a.

Mikhelson IV, Lee P, Bakhtiari S, Elmer TW, Katsaggelos AK, Sahakian AV. Noncontact millimeter-wave real-time detection and tracking of heart rate on an ambulatory subject. IEEE Trans Inf Technol Biomed 2012b;16:927–934.

Moghaddar A, Walton EK. Time-frequency-distribution analysis of scattering from waveguide cavities. IEEE Trans Antennas Propag 1993;41(5):677–680.

Möller HG. Über Störungsfreien Gleichstromempfang mit den Schwingaudion. Jahr fur Draht Teleg 1921;17:256–287 (note: See the citation in Adler (1946) to confirm special characters in German. Google Translate suggests Über Störungsfreien Gleichstrom Empfang mit den Schwingaudion which means On Fault-free DC reception with the oscillating audion in English.).

Morinaga M., Nagasaku T., Shinoda H., H. Kondoh. 24 GHz intruder detection radar with beam-switched area coverage, IEEE MTT-S International Microwave Symposium; Honolulu (HI); 2007 Jun. p. 389–392.

Mostafanezhad I, Park B, Boric-Lubecke O, Lubecke V, Host-Madsen A. Sensor nodes for Doppler radar measurements of life signs. IEEE MTT-S Int Microw Symp DigHonolulu (HI) 2007:1241–1244.

Mostafanezhad I, Boric-Lubecke O. An RF based analog linear demodulator. IEEE Microw Wireless Compon Lett 2011;21(7):392–394.

Mostov K, Liptsen E, Boutchko R. Medical applications of shortwave FM radar: remote monitoring of cardiac and respiratory motion. Med Phys 2010;37(3):1332–1338.

National Institute of Health News. Available at http://www.nih.gov/news/health/apr2010/nhlbi-08.htm. Accessed 2013 May 26.

Nguyen T-K, Krizhanovskii V, Lee J, Han S-K, Lee S-G, Kim N-S, Pyo C. A low-power RF direct-conversion receiver/transmitter for 2.4-GHz-band IEEE 802.15.4 standard in 0.18-μm CMOS technology. IEEE Trans Microw Theory Tech 2006;54(12):4062–4071.

Obeid D, Sadek S, Zaharia G, El Zein G. Cardiopulmonary activity monitoring with contactless microwave sensor, Mediterranean Microwave Symposium; Istanbul, Turkey; 2012.

Paciorek LJ. Injection locking of oscillators. Proc IEEE 1965;53(11): 1723–1727.

Page RM. The early history of radar. Proc IRE 1962;50(5):1232–1236.

Pan W, Wang J, Huangfu J, Li C, Ran L. Null point elimination using RF phase shifter in continuous wave Doppler radar system. Electron Lett 2011; 47(21):1196–1198.

Park B, Boric-Lubecke O, Lubecke VM. Arctangent demodulation with DC offset compensation in quadrature Doppler radar receiver systems. IEEE Trans Microw Theory Tech 2007;55:1073–1079.

Park Z, Li C, Lin J. A broadband microstrip antenna with improved gain for non-contact vital sign radar detection. IEEE Antennas Wireless Propag Lett 2009;(8):939–942.

Petkie DT, Benton C, Bryan E. Millimeter wave radar for remote measurement of vital signs, IEEE Radar Conference; Pasadena (CA); 2009. p. 1–3.

Petrochilos N, Rezk M, Host-Madsen A, Lubecke V, Boric-Lubecke O. Blind separation of human heartbeats and respiration by the use of a Doppler radar remote sensing. IEEE ICASSP 2007Honolulu (HI) 2007.

Rao SS. *Mechanical Vibrations*. Reading (MA): Addison-Wesley; 1986. ch. 3.

Rasshofer RH, Biebl EM. Advanced millimeter-wave speed sensing system based on low-cost active integrated antennas. IEEE MTT-S Int Microw Symp 1999;1:285–288.

Razavi B. A study of injection locking and pulling in oscillators. IEEE J Solid-State Circuits 2004;39(9):1415–1424.

Reyes G, Wang D, et al. VitalTrack: a Doppler radar sensor platform for monitoring activity levels, IEEE Radio and Wireless Week; Santa Clara (CA); 2012 Jan 15–18.

Rice JA, Li C, Gu C, Hernandez J. A wireless multifunctional radar-based displacement sensor for structural health monitoring. In: *SPIE Smart Structures/NDE*. San Diego (CA): 2011.

Rice JA, Gu C, Li C, Guan S. *A Radar-Based Sensor Network for Bridge Displacement Measurements*. San Diego (CA): SPIE Smart Structures/NDE; 2012.

Richards MA. *Fundamentals of Radar Signal Processing*. McGraw-Hill Electronic Engineering, New York, NY, USA; 2005.

Shahid N, Fang D, Sheng W, Ye X. The high resolution estimate—a comparative study, Proceeding of IEEE ICCEA, 1999, p. 262–265.

Shipton HW. Radar history: the need for objectivity. IEEE Trans Aerosp Electron Syst 1980;AES-16(2):44.

Singh A, Lubecke V. Respiratory monitoring using a doppler radar with passive harmonic tags to reduce interference from environmental clutter. EMBC:2009 Annual International Conference of the IEEE Engineering in Medicine and Biology Society, 2009:3837–3840.

Singh A, Lubecke VM. "Respiratory Monitoring and Clutter Rejection Using a CW Doppler Radar With Passive RF Tags," IEEE Sensors Journal 2012;12(3):558–565.

Silva CW. *Vibration*: Fundamentals and Practice. 2nd ed. CRC Press, Taylor & Francis Group, Boca Raton, FL, USA; 2007.

Staderini EM. UWB radars in medicine. IEEE Aeros Electron Syst Mag 2002: 13–18.

Stezer A, Diskus CG, Lubke K, Thim HW. Microwave position sensor with sub millimeter accuracy. IEEE Trans Microw Theory Tech 1999;47(12): 2621–2624.

Stockbroeckx B, Vorst AV. Copolar and cross-polar radiation of Vivaldi antenna on dielectric substrate. IEEE Trans Antennas Propag 2000;48(1): 19–25.

Stove AG. Linear FMCW radar techniques. IEE Proc Radar Signal Proc 1992;139:343–350.

Swords SS. *Technical History of the Beginnings of Radar*. The Institution of Engineering and Technology (IET), UK; 1986.

Tahmoush D, Silvious J. Remote detection of humans and animals, IEEE Applied Imagery Pattern Recognition Workshop (AIPRW); Washington (DC); 2009 Oct 14–16.

Trintinalia L, Ling H. Interpretation of scattering phenomenology in slotted waveguide structures via time-frequency processing. IEEE Trans Antennas Propag 1995;43(11):1253–1261.

Trintinalia L, Ling H. Joint time-frequency ISAR using adaptive processing. IEEE Trans Antennas Propag 1997;45(2):221–227.

Wang F-K, Li C-J, Hsiao C-H, Horng T-S, Lin J, Peng K-C, Jau J-K, Li J-Y, Chen C-C. A novel vital-sign sensor based on a self-injection-locked oscillator. IEEE Trans Microw Theory Tech 2010;58(12):4112–4120.

Wang F-K, Horng T-S, Lin J, Peng K-C, Jau J-K, Li J-Y, Chen C-C. Single-antenna Doppler radars using self and mutual injection locking for vital sign detection with random body movement cancellation. IEEE Trans Microw Theory Tech 2011;59(12):3577–3586.

Watson RC Jr,. *Radar Origins Worldwide: History of Its Evolution in 13 Nations Through World War II*. Trafford Publishing, Victoria, BC, Canada; 2009.

Xiao Y, Lin J, Boric-Lubecke O, Lubecke VM. Frequency tuning technique for remote detection of heartbeat and respiration using low-power double-sideband transmission in Ka-band. IEEE Trans Microw Theory Tech 2006;54:2023–2032.

Yan H, Yin F, Zhu G, Ajlouni M, Kim J. The correlation evaluation of a tumor tracking system using multiple external markers. Med Phys 2006;33(11):4073–4084.

Yan Y, Li C, Yu X, Weiss MD, Lin J. Verification of a non-contact vital sign monitoring system using an infant simulator, EMBC: 2009 Annual International Conference of the Ieee Engineering in Medicine and Biology Society; 2009. p. 4836–4839.

Yan Y, Li C, Lin J. Effects of I/Q mismatch on measurement of periodic movement using a Doppler radar sensor, Proceedings of IEEE Radio and Wireless Symposium; New Orleans (LA); 2010 Jan 10–14.

Yan Y, Li C, Rice J, Lin J. Wavelength division sensing RF vibrometer, IEEE MTT-S International Microwave Symposium; Baltimore (MD); 2011a Jun.

Yan Y, Cattafesta L, Li C, Lin J. Analysis of detection methods and realization of a real-time monitoring RF vibrometer. IEEE Trans Microw Theory Tech 2011b;59(12):3556–3566.

Yang Y, Wang Z, Fathy AE. Design of compact Vivaldi antenna arrays for UWB see through wall applications. Prog Electromagn Res 2008; 82:401–418.

Yang Y, Fathy A. Development and implementation of a real-time see-through-wall radar system based on FPGA. IEEE Trans Geosci Remote Sens 2009;47(5):1270–1280.

Zhang C, Fathy AE. Reconfigurable pico-pulse generator for UWB applications, Proceedings of the Microwave Symposium Digest; 2006 Jun. p. 407–410.

Zhao X, Song C, Lubecke OB, Lubecke VM. DC coupled Doppler radar physiological monitor, 33rd Annual International Conference of the IEEE EMBS; Boston, Massachusetts (MA); 2011 Aug.

Zhou Q, Liu J, Host-Madsen A, Boric-Lubecke O, Lubecke V. Detection of multiple heartbeats using Doppler radar. IEEE ICASSP 2006 Proc 2006;2:1160–1163.

INDEX

Microwave Noncontact Motion Sensing and Analysis, First Edition.
Changzhi Li and Jenshan Lin.
© 2014 John Wiley & Sons, Inc. Published 2014 by John Wiley & Sons, Inc.

WILEY SERIES IN MICROWAVE AND OPTICAL ENGINEERING

KAI CHANG, Editor
Texas A&M University